'Perceptive and evocative prose ... This is the most insigh... nating book about the King of Beasts and its world that I have read.'

—Dr. George Schaller, author of *The Serengeti Lion* and travel companion to Peter Matthiessen in *The Snow Leopard*

'This is the first book I have read that does justice to the magnificent and imperilled existence of the African lion. Anthony Ham has dug deeply beyond the headlines to unravel the complexity of our relationship with Africa's great cat, and explain the formidable conservation challenges it faces today. In tense, evocative and often heartbreaking prose, he captures the blood and dust of the lion's life. And by immersing himself in the lives of those people who walk in their paw-prints, he produces sympathetic and often revelatory insight. This is lyrical and gripping storytelling of authentic Africa.'

—Luke Hunter, Executive Director, Big Cats Program, Wildlife Conservation Society (WCS), and author of *Wild Cats of the World* and *Cats of Africa*

'Anthony Ham is reporting from the edge of oblivion. In this moment of rapid and epochal change, these hard-won tales of lions and the humans who compete with them read like the last transmissions from a fast-receding world. Bravely pursued, acutely observed, and elegantly told, these true stories of the human–lion endgame may, like the ancient beings and cultures they so compellingly describe, be the last of their kind.'

—John Vaillant, bestselling author of *The Tiger*

'Urgent and important. This moving tale with a heroic cast of characters, leonine and human, is a must-read for anyone passionate about wildlife and wild places.'

—Tony Park, bestselling author of *Last Survivor*

For Carlota and Valentina

THE LAST
LIONS OF AFRICA

Stories from the frontline
in the battle to save a species

ANTHONY HAM

ALLEN&UNWIN
SYDNEY · MELBOURNE · AUCKLAND · LONDON

Allen & Unwin
83 Alexander Street
Crows Nest NSW 2065
Australia
Phone: (61 2) 8425 0100
Email: info@allenandunwin.com
Web: www.allenandunwin.com

A catalogue record for this
book is available from the
National Library of Australia

ISBN 978 1 76087 575 6

Maps by Guy Holt Illustration and Design (guyholt.com)
Internal design by Post Pre-press Group
Set in 12.5/17.5 pt Minion Regular by Post Pre-press Group, Australia
Printed and bound in Australia by The SOS Print + Media Group

10 9 8 7 6 5 4

Contents

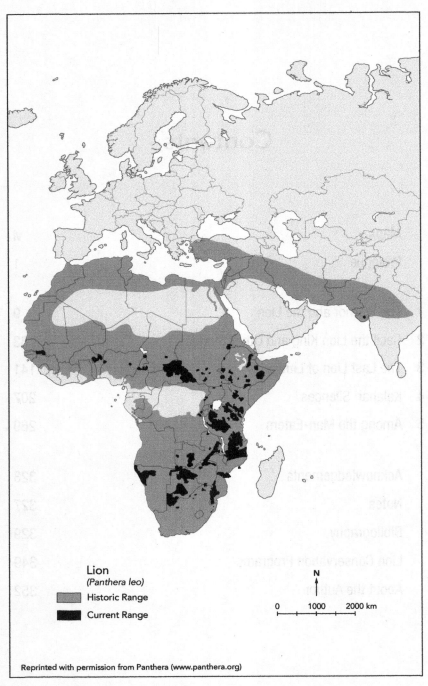

Lion
(Panthera leo)

■ Historic Range

■ Current Range

N

0   1000   2000 km

Lion distribution, historic and current

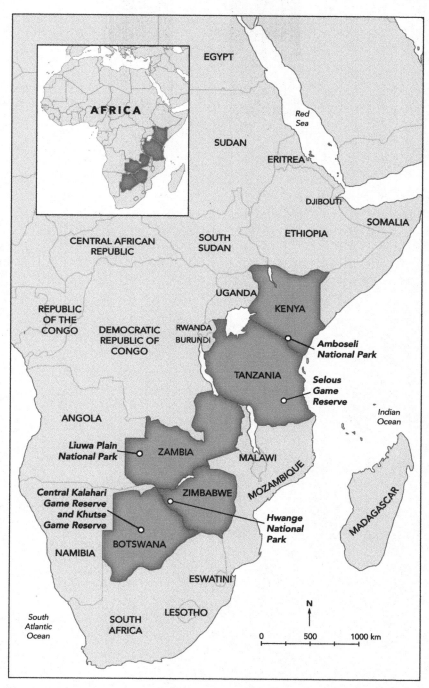

Countries and national parks covered in this book

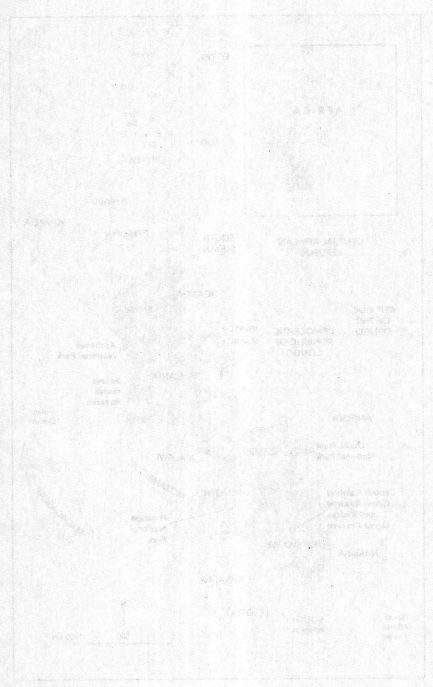

Countries and national parks covered in this book

Prologue

Not so long ago, along the boundary of Hwange National Park in Zimbabwe, there lived a troubled lioness. On one side, a crowded lionscape of prides inside the park made her life miserable. On the other, people and villages pressed close up against the park boundary. To the west, she could be killed by other lions; to the east, she risked the wrath of human beings every time she ate one of their cows. The lioness took what she could, just to survive, leaving the park authorities with two choices: kill the lion—a solution known as Problem Animal Control (PAC)—or move her elsewhere. They chose the latter. It seemed like a good idea at the time, and animal welfare advocates cheered. In an age where lions are in trouble, saving every lion matters.

If only it were that simple.

Having captured the lioness, no one quite knew what to do with her next. The reason the lioness was out in the world of people was because there was no room left for her in the world of lions. Hwange National Park is big, but it's nothing compared to where lions used to live. Every corner of the national park has its own pride of resident lions and none of them take kindly to intruders. Translocate her into one of these areas and she either gets killed or she's forced from the park—and kills more cows. 'It's very unfair to go and throw a lion on top of others, for the one that's being dumped *and* the ones that are being dumped on,' said one lion researcher from Hwange. 'They must be thinking, "Where the bloody hell did that thing come from?" And then everybody starts fighting.' And besides, if you don't want a lion to return, you must move it at least 300 kilometres away from where it once lived. Otherwise it finds its way home.

In the end they took her to Kazuma, an area north-west of the park, where they placed her in an enclosure to allow her to grow used to her new surroundings. It wasn't an ideal solution. To save this one stressed lioness, captured and caged, other animals had to be killed in order to provide her with food. When they finally released her, the same old problems resurfaced. Kazuma is close to Zimbabwe's boundaries with Botswana and Zambia, between the towns of Pandamatenga and Kazungula. On Zimbabwe's side of the border, there are farms and hunting concessions (privately owned land where hunting is permitted), and the park itself isn't far away.

The lioness survived for a time. Pressured at every turn by people and by other lions, she drifted north and crossed the Zambezi River into Zambia. There she found a male with whom she mated and had cubs. But that corner of Zambia is even more densely populated with people than where she'd come from, and wild prey is largely non-existent. To feed her cubs, just to stay alive, she again began killing cattle. For her crimes she was shot and killed. Her cubs were caught and placed in captivity, where they remain to this day.

~

Do we have room in our world for lions? In 2011, haunted by the idea that lions may disappear from the planet in my children's lifetimes—if not my own—I embarked on a series of journeys across East and southern Africa, looking for stories that might answer that question. Driven by a passion for wilderness and wild places, I brought back many stories. The five that tell us the most about lions are in this book.

On these journeys, I learned that lions are in trouble. If the experience of the Kazuma lioness teaches us anything, it is that we still have much to learn if lions are to have a future. In 2019, the most detailed study of lions on the African continent estimated that 22,509 lions remained. That may sound like a lot of lions—there are fewer than 4000 wild tigers left in the world and only around 1000 mountain gorillas. Yet at the end of the nineteenth century, there were probably 200,000 lions in Africa. Lions have disappeared from 95 per cent of their historical range and from 26 African countries. Of those lions that survive, many live in isolated and fragmented populations that have little hope of long-term survival.

It is impossible to tell the story of lions without also venturing into the world of those who live alongside them. It is easy to talk of saving lions from a distance. It is much more complicated when you're a subsistence farmer and

3

a lion eats your livestock or, worse, your children. Lions inhabit the stories of Africa's traditional peoples, often as symbols of royalty and power and rarely as enemies. They are sometimes feared but almost always respected. Now, at a time when traditions are dying and lions are disappearing, people and lions face a desperate struggle for survival, increasingly in conflict with one another in battles that no one wins. As human populations grow across Africa—by one United Nations estimate, Africa's population will triple in the 50 years to 2050, climbing above 2.5 billion people— lions and people will live in ever closer proximity out upon the savannah of East Africa and in the deserts of the south, by the waterholes that lions and people depend upon, and along rivers where hungry lions lie in wait.

The plight of lions is also a warning signal that Africa's wilderness areas are in trouble. Lions are our finger in the dyke, the keystone species holding it all together. Take lions out of the equation and ecosystems fall apart. Without lions, populations of other species increase unchecked, habitats are destroyed, and there is no barrier to humankind's final destruction and desperate takeover of barely habitable land. The disappearance of lions is the beginning of the end for Africa's last wild places, the final step in the irreversible decline and death of so much African wilderness. Lions, and possibly only lions, may be the saviours of Africa as we know it.

These three enduring African characters—lions, the traditional peoples they live among, and the wild lands of Africa that together they inhabit—are what this book is about. And for every glimmer of hope that these stories offer, there is always a hint of elegy, a lament for all that has been lost or may soon be gone.

~

The first time I saw wild lions, I felt as if I had snatched a glimpse of eternity. These lions, truly wild, stalked the earth as if this were their time, as if the world could forever be like this. There is nothing quite like a wild lion to evoke that shudder of fear, that frisson of excitement that animates the world's last wild places. Enduring symbol of kings, possessed of terrible heraldic beauty, lions inhabit the darkest recesses of our psyche, like the fearsome monsters who once stood in for terrifying mysteries and our longing for exploration on maps of the ancient world. In the leonine profile, savage and elusive, there resides a deliciously untamed life force that inspires the same awe that we reserve for the great mountain ranges of the earth, for ocean storms, and for epic sand-dune seas. This is what drew me to lions.

The stories in this book follow characters: lions with real names like Cecil, Nosieki and Lady Liuwa; people like Meiteranga the Maasai warrior and Induna Mundandwe the Barotse chief. It's what happened when a Maasai warrior in Kenya killed a lion only to find himself faced with an existential crisis that threatened everything he and his people stood for. It's what really happened to Cecil, who showed us that even lions deep within national parks may not be safe. One story chronicles the life of Lady Liuwa, the last lioness of western Zambia who became a goddess to the local people, and whose story is a reminder that when lions disappear from a land, it can be terribly difficult to bring them back. Another traces a solo crossing of the Kalahari in Botswana, searching for the last survivors in a land emptied of people and of lions. Stories from south-eastern Tanzania chart a world under the spell of Africa's most prolific man-eating lions, cautioning us as to what happens when things fall apart. And I tell of my own near-death experience with a lion in the African dawn.

~

Time is running out for lions. An oft-recounted Nairobi legend, no longer verifiable, holds that the first six people to be buried in Nairobi's cemetery were killed by lions. Back then, in the final days of the nineteenth century, Africa belonged to the lions.

Little more than a century later, in June 2012, six lions stepped beyond the boundaries of Nairobi National Park and into Kitengela, where a group of angry local Maasai cornered them, eager to exact revenge for the loss of their livestock to predators. Or perhaps they were angry because the intrusion of lions into the human world was an outrage, a threat that could not go unanswered. And who could blame them? Facing lions on suburban streets should not be one of the more routine challenges of daily life. 'Some people think having lions in the neighbourhood is the best thing ever,' one Nairobi resident told London's *Times* newspaper. 'But I don't want them to eat my kids.' The mob speared the six lions to death before help could arrive.

Prologue

There is a certain symmetry to these Nairobi tales of death—a story of six lions, the other a story of the first six people to be buried in the city's cemetery, all killed by lions. Such has been the relationship between lions and people in Africa for the past hundred years. The balance has shifted. The lions no longer win.

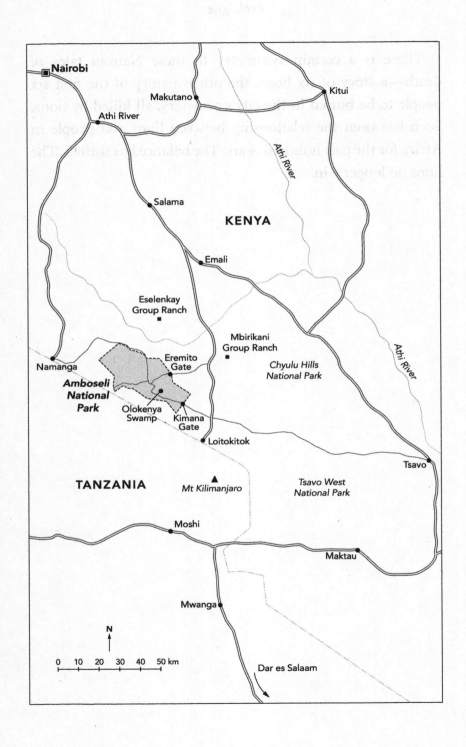

1

The Warrior and the Lion

Kenya, 2011¹

When Meiteranga Kamunu Saitoti killed his first lion, he hoped it was only the beginning.

He was just nineteen years old back then and had dreamed of killing lions for as long as he could remember. How he came to do so echoes Africa's oldest stories. A lion had killed one of the Maasai's prized cows out on the plains of Amboseli in southern Kenya and, eager for revenge, young Maasai men came together in the gathering gloom of late afternoon. Tempers flared—some wanted to go immediately, others thought it wiser to wait until morning. As word spread, more young men came in from the wild to join the hunt, and with more chaos than careful planning they formed a hunting party of earnest young men baying for blood.

Bristling with the hormones of a young Maasai warrior-in-waiting, Kamunu, as he was then known, could barely restrain his excitement. Destiny called to him down through the generations—his father and his uncles killed fifteen lions between them—and, with the impatience of youth, he longed to be like them; everything in his life until then had prepared him for this moment. But killing the lion would not be enough. He had to throw the first spear.

Night fell, and the lions disappeared into thick thorn scrub where no Maasai would follow in darkness.

The young Maasai talked into the night. Perhaps if there had been only one or two of them, their fury may have subsided. Perhaps. But with so many of them to stoke the fire in the uncertainty of the night, their anger never cooled. Kamunu, who always stood apart from his peers, tried to remain aloof and to sleep, but he could do neither.

If anyone was afraid they dared not show it, and when morning arrived eleven young men carrying spears drew near to one another, seeking courage and warmth in the dawn chill. To anyone watching, save for the envious Maasai children and admiring young Maasai women who paused on their morning march for water, these pigtailed warriors setting out in pursuit of lions must have risen from the earth like apparitions, an echo of an archaic Africa long thought extinct. Cloaked in *shukas*—thin cotton blankets in black and red—they seeped like blood out across the land.

From Kamunu's youngest days, the lion-killers in his life had regaled him with tales of their heroics. One of his earliest memories was of being a little boy out on the plains, herding his family's cows, when a lioness began stalking the herd. He screamed, the dogs barked, and the lioness retreated into the bush. Years later, Kamunu would remember two things about that first encounter: that the lioness was 'very beautiful', and that he longed for the day when he could kill her and her kind.

As a young boy, he learned how to track wildlife through the bush. Was it lion or leopard? If lion, male or female? How many were there? How long ago had they passed by? In which direction were they travelling? In time, like well-trained scientists in the field, most Maasai can tell individual lions apart. Such knowledge might come from the frequency with which they observe a lion in a known territory. Perhaps they can read the distinguishing marks on a lion's body, or a lion's behavioural quirks. The real experts among them may, when they get close enough, even be able to distinguish lions by the configuration of whisker spots arrayed around the muzzle. Whatever their method, young Maasai learn to identify the lions with which they may one day be forced to do battle—a skill as invaluable to the young herder of cattle charged with keeping his family's livestock safe as it is to the would-be warrior who will one day follow lions through the bush in order to hunt them.

Through it all, history and expectation had weighed upon Kamunu. To become a Maasai warrior—no easy task—was something that every young Maasai man dreamed about. Certain rituals had to be fulfilled. Like generations of Maasai before him, Kamunu had submitted to the senior-boy ceremony, wandering throughout the Amboseli Basin with other precocious teenagers to announce his presence as a warrior-in-waiting. Circumcision then marked the threshold into manhood. And then, everything in its turn, Kamunu entered the *emanyatta*, a warrior's camp, as generations of young Maasai men had before him; there he learned the complex ways of warrior-craft. Only with this initiation complete did he emerge as a *murran*, or true warrior.

Even then, young Maasai warriors still had to prove their bravery, their readiness to join the storied ranks of warriors past. They did this by killing a lion. This rite of passage, the point at which the fates of lions and the Maasai become entangled in a common destiny, is known as *olamaiyo* in the Maa tongue.

On Kamunu's first hunt, when he was nineteen years old and had only recently emerged as a *murran*, the would-be warriors were at first in high spirits: they stood, after all, on the cusp of what could become the defining moment of their lives. But as the day wore on, the lions—an adult female, an adult male and two cubs—continued on the run. Their Maasai pursuers tired. Some wandered off. The dwindling party followed the lions until just a handful of Maasai remained. By late afternoon, they were still going.

'The rest of the group was getting tired,' Kamunu recalled years later. 'We had been going all day. The lions, too, were tired because the day was hot. I wanted to take advantage of that tiredness in the others.'

As he told me the story, close to where the hunt had taken place twelve years earlier, the air was dry as tinder and crickets whirred in the late heat of a day, not unlike the one when Kamunu and his Maasai friends stalked four lions through the bush. I tasted dust on my tongue. On such days, in such places, the air thrums without pause; it is so constant that you wonder whether, even in utter stillness, what you can hear is the sound

of the Earth turning. In the torpor of the East African afternoon, amid a degraded landscape of sand and thorn brush,
such vivid tales of lions and warriors belonged to an altogether
more epic time and place.

'The lioness, she knew we were coming,' Kamunu told me.
'I went to the tree where I thought she was. But she wasn't there.
Suddenly, there she was and she charged. She ran straight at me,
trying to protect her cubs. I threw the spear. The spear hit her
in the shoulder. She went down. I jumped out of the way. Then
someone else threw his spear. And then everyone joined in.'

The lioness didn't stand a chance.

He took home the lion's tail and, like all Maasai who kill
a lion, a new name. Kamunu was now Meiteranga, 'The One
Who Was First', reborn as the killer of lions.

There is something primordial about the scene that
followed: colourfully dressed warriors dancing around, then
dismembering the still-warm body of a lion upon East Africa's
parched carapace. As is the Maasai way, the celebrations and
the dancing went on for days, and Meiteranga was at the centre
of it as he moved easily through the festivities, carried along
on a tide of emotion and wellbeing, the undisputed king of
his world as his peers paid homage; it was his finest moment.
A lion was dead, and Maasai culture—at least in those joyous
moments—was very much alive. Meiteranga glowed with
pride, basking in the adulation of family and friends as he
paraded the lioness's tail for all to see. And when it all came to
an end, he slept, and he did so soundly.

Later, when I came to know him, Meiteranga radiated
certainty and one suspects that it was on this day, the day when

13

he killed his first lion, that this certainty first took hold. Life for Meiteranga was black and white. He knew where he stood and it was at the head of a long line of lion-killers. Meiteranga is not the kind of man who doubts.

The story could have ended there, for the lions and for Meiteranga. But, of course, it didn't. And the key to understanding what happened later lies in a simple detail from the scene, a detail that Meiteranga hardly noticed at the time: the cubs of the slain lioness escaped into the bush. One of those cubs would one day return to haunt him.

~

In 2011 I travelled to Nairobi, my gateway into the world of lions. Along the road into town from the airport, drivers looked away and in the gloom it might have been any large city of questionable reputation—sagging roofs of corrugated iron; cowled figures crouched beneath thin shelters; fires in barrels; sallow, pinch-faced dogs; empty lots; hostile shouts; the stench of fruit left to rot. Beyond, all was darkness. Nairobi was a city of shadows and silhouettes.

When day came, steam rose from mounds of rubbish by the roadside under a bright October sun, while men in dark suits and women in high heels picked their way around muddy potholes sheened with oil; many of these foot commuters had abandoned gridlocked traffic. Horns and petrol fumes, shouts and hands thrown high in exasperation: Nairobi for much of the day roiled in perpetual noise, a city on the move without ever seeming to go anywhere.

Safari touts drew near to my open window. I shut out their chatter and listened instead to phone-in radio shows on which local residents railed against their city's problems, all the while professing it unimaginable that they should live anywhere but this, the greatest city on earth.

This greatest city on earth began, like so many African cities, on a colonial whim. Until the last decade of the nineteenth century, the Maasai inhabited this land and only a series of swamps distinguished it from the surrounding savannah; the Maasai called it *uaso or enkare nyrobi* (cold water). That the British chose this site and not, for example, the fertile soils of the Rift Valley owed much to this presence of water. The swamps also marked the approximate midpoint of the Mombasa–Kampala Railway, a critical tool in opening up the East African interior to colonial expansion. Nairobi, a suitable distance from the well-provisioned settlements along the coast, served as a railway supply depot and a statement of intent for a colony determined to make East Africa its own.

With the Maasai duly dispatched, the new city, founded in 1899, was a rugged, poorly protected place peopled by those tough frontier types—ex-soldiers and soldiers of fortune, ambitious farmers and rogues with little left to lose—drawn to the outer reaches of empire. Nairobi's high elevation—the city rises some 1660 metres above sea level—and cooler climate also attracted colonial administrators and their families desperate to escape the sweltering heat and malarial humidity of the coast. In very short order, Nairobi grew to become the premier colonial city in the region. By 1907 it was the capital of British East Africa.

For decades after the city's founding, lions and ill-tempered rhinoceroses infiltrated Nairobi's streets. The records are silent as to whether the first six unfortunates to be buried in the city's cemetery were ruddy, hard-bitten adventurers who died in mortal combat with lions, or wives of public servants taken on their way to dinner parties in neck-to-ankle dresses. The authorities pleaded with residents to carry guns whenever they left their homes. Others needed no invitation.

Nairobi's proximity to some of the richest wildlife concentrations on the planet lured swaggering white hunters to the city. Six people may have been eaten by lions when the city was born, but in time hundreds, perhaps thousands more lions were killed in the name of progress and of sport. Feeding on this frenzy, Kenya's first hotels appeared; the once-palatial Norfolk Hotel, built in 1904, still stands. Even so, the city retained its rough edges, and the calamities to which Nairobi remained vulnerable—an outbreak of plague, a fire that effectively destroyed the town—soon filled the cemetery with hundreds of victims who were most definitely not killed by lions. In 1926, the private secretary to the British governor of Kenya, Eric Dutton, described Nairobi as 'a slatternly creature, unfit to queen it over so lovely a country'.

White East Africans are known for yearning—openly, and uncomfortably within earshot of black Kenyans—for the city's former aura of colonial romance, for the time when the great white hunters, Ernest Hemingway among them, could hold court in the bars of stately old hotels. But that city long ago ceased to exist. While I was in Nairobi, renovations closed the Lord Delamere Terrace at the Norfolk Hotel for the first time in its history. However necessary those renovations may have been, they closed the door on a storied past, and marked the end of something.

Modern Nairobi, while not without charm or pockets of affluence, seemed a place to transact essential business, to organise journeys out into the wild, and for passers-through to linger no longer than necessary. For its African inhabitants it was very often another thing altogether. In the early 1900s just 10,000 people lived in Nairobi; 3.5 million now call it home. Yet just as it did when Dutton visited in 1926, inadequate infrastructure blighted Nairobi at the time of my visit, and its inhabitants very often lived in appalling conditions.

The longer I stayed, the less I understood about this brash and thoroughly African city. In the traffic-choked streets east of Moi Avenue, after taking two hours to travel less than a kilometre, I left Peter Ndirangu, my driver, by the roadside to set off on an errand. When I returned a few minutes later, three plain-clothes policemen occupied the car, having arrested Peter and impounded his vehicle. He had, it seemed, parked illegally. Money would later change hands and our misdemeanour would be forgiven. But when I pointed out to one of the officers the absurdity of being charged with parking illegally when the surrounding traffic had not moved in half an hour, he regarded me with disdain: 'I'm sorry, sir, but you do not understand. You are a foreigner.'

By night I slept badly, haunted by the marabou storks glowering, grim reaper–like, over the immobile traffic; each morning I woke to the mournful cries of the hadada ibis. That

I could hear a wild animal of any description there should have been cause for celebration. Instead, the birds' presence caused disquiet, a feeling that this was all the wilderness that remained: bald-headed birds presiding over the fuming traffic jams and fetid rubbish dumps of the city.

~

From first light, clouds hung low over Nairobi like messengers of doom, and occasional downpours did nothing to relieve the suspense. People looked skyward, worrying, even in the city, that the rains would be late again. Memories of 2009 were still fresh, when the worst drought in living memory brought starvation and tens of thousands more inhabitants to the city's slums, as well as the fear that a reckoning would one day come. In the meantime, everyone prayed for rain.

On one such overcast Nairobi day I escaped to the august surrounds of Nairobi National Museum, one of the great colonial institutions in East Africa. There I was to meet Dr Darcy Ogada, a conservation biologist originally from upstate New York. For Dr Ogada, the most serious threat to Kenya's wildlife was the use of poisons, primarily pesticides and most commonly a granular pesticide called carbofuran, to wipe out predators.

'I haven't heard of any wildlife poisonings for months,' Ogada began as soon as we sat down, 'and then just this morning I received an email reporting that two lions—a lioness and her cub—were poisoned near Amboseli.'

In the past two decades, Ogada told me, the conflict between humans and wildlife had grown exponentially in Kenya. 'The poisoning of wildlife here is rampant,' she explained. 'For herders, it's usually about lions. For farmers, it's everything. These farmers are far more tolerant of pests and threats than Western farmers will ever be, but these people eke out a subsistence existence and many have now reached breaking point. The response from the wildlife authorities has been dismal. People are just desperate.'

Before I could ask a question, Ogada continued, barely pausing for breath.

'These farmers and herders don't have a history of killing wildlife. Talk to the elders—they respect the animals. It's a part of their heritage. The younger generation—that's where the problem lies. They've lost their connection with their culture, with the land, and with the wild animals.'

The poisoning of lions—53 killed in the Laikipia region of northern Kenya in the decade from 2001, 68 around Amboseli during the same period—had, by the time of my visit, reached levels that threatened the existence of all lions in Kenya. And it wasn't only lions under threat. When predators are finished with a kill, the scavengers move in to feed on the carcass, and hyenas, too, had been dying in record numbers. For vultures, Ogada said, there have been 'staggering declines in abundance'. When it came to lions, she continued, 'There can be little doubt that these reported incidents represent just the tip of the iceberg.'

The international spotlight shone briefly upon the poisoning of Kenya's lions when CBS's *60 Minutes* aired a damning report in 2009. In response, FMC, the largest supplier of carbofuran pesticides, withdrew its products from the Kenyan market. But the killing continued, and the publicity, if anything, made life more difficult for those involved in the struggle. Ogada was the only activist who was willing to go on record to criticise the powerful agricultural lobby that fought any attempts to take poisons and pesticides off the market.

One activist asked to remain anonymous and begged me not to name the organisations in question. Another high-profile Kenyan agreed to meet with me and then later reneged, explaining that 'I am on the front line, and as you can imagine this is a hugely sensitive issue with massive implications for me and for Kenya'. Ogada repeated rumours that the member of Kenya's parliament who raised the issue in the aftermath of the *60 Minutes* furore was 'silenced'.

'It's never going to stop,' she told me as we parted, 'not until all of the predators are dead.'

~

Soon after our conversation, I fled to Nairobi National Park, eager to see lions. But fences and suburbs encircled the park's zebra and wildebeest, which could no longer move with the seasons and migrate with the rains. The park's predators, too, could no longer escape, hunted and killed if they strayed beyond the park in search of new territories, or doomed to mate with their offspring and face the consequences of a stagnant gene pool if they remained. The tall buildings of Nairobi's central business district loomed on the near horizon. The park felt like a zoo.

As fast as Nairobi's traffic would allow, I hurried out onto the Athi-Kapiti Plains that sweep away south-east of the city. In Nairobi's early years, the wildlife concentrations there rivalled those of Tanzania's Serengeti and the Masai Mara in Kenya's south-west, and the plains were favoured hunting grounds for the great white hunters of the day, among them Theodore Roosevelt, Denys Finch Hatton and Abel Chapman. Back then, when Nairobi's population numbered in the tens of thousands, the plains were 'swarming with gnu', according to one chronicler of the day.

Until well into the twentieth century, these plains witnessed one of Africa's greatest migrations as zebra moved unimpeded between what is now Nairobi National Park and the foothills of Mount Kenya in the north, or Amboseli hundreds of kilometres away to the south-east. Wildebeest, too, travelled here en masse from as far as Tsavo and the Chyulu Hills, some 250 kilometres away, to calve. As late as 1961, a pride of 40 lions lived on the plains.

But then the door began to close as Nairobi's human population grew.

At first, a patchwork of fences appeared across the National Park's southern hinterland. Having begun slowly, the growth in human settlements gathered quiet momentum until the 1990s, when an export processing plant was established at Kitengela, close to the southern fringes of the park. The last corridors of uninhabited land narrowed and then disappeared. To cross from the park to the Athi-Kapiti Plains, wildlife had to run a gauntlet of people, fences and domestic animals.

Between 1977 and 2002, wildlife populations on the Athi-Kapiti Plains fell by 72 per cent. Numbers of zebra, eland and wildebeest crashed by almost 90 per cent during the same period. At the start of the twentieth century, there were four wild animals for every head of livestock on the Athi-Kapiti Plains. Now it is the other way around. In April 2004, near the burgeoning settlement of Athi River, someone laced cow carcasses with poison. When they came to feed, 187 vultures died and the poison wiped out most of the local hyena population in a single day.

With Nairobi visible away on the horizon, I dropped off the Nairobi–Mombasa highway, seeking refuge from dark thoughts and dire predictions in the Swara Plains Conservancy. Straggling herds of wildebeest, giraffe and Thomson's gazelle picked at meagre grasses within sight of a newly built cement factory, its sign bearing the image of a male lion ('Simba Cement: King of the Concrete Jungle'). The factory belched smoke into the air; the fenced landscape of rolling plains had already turned the corner into domesticated land. Even so, there were moments when the open horizons, waterholes and wandering giraffe carried echoes of a more abundant past, bearing on the light wind suggestions of Kenya's golden age of wildlife and wild places.

Close to the conservancy's eastern limits, I enjoyed a sundowner with Sandy Simpson, a softly spoken white Kenyan. Anger in Sandy ran just below the surface. He railed against the influential NGO Friends of Nairobi National Park for making pronouncements about the need to protect the park's wildlife corridors, yet doing so from their large offices in downtown

Nairobi. 'Why don't they sell their offices and buy up land along those corridors?' he asked. It seemed like a fair question.

After leaving behind a lucrative jewellery business in Europe, Sandy began whittling away his life savings in noble but doomed causes. Above all he dreamed of uniting the landowners of Athi-Kapiti, of creating a conservation project that restored the plains to their former glory. At the same time he found himself caring for five male lions. Refugees from Nairobi National Park, these lions had to be rescued from themselves—when they left the park to try to establish their own territories, a wall of human settlements blocked their path. Considered a threat to life and livestock, the lions could no longer range freely in the wild and Sandy hadn't the heart to put them down. It was clear that they were still wild—they lunged at him and at me when we strayed too close to the fence.

Both of Sandy's projects were hot, unforgiving work. He already despaired, he confessed, of ever finding a unity of will among the local landowners to save what remained of the ecosystem.

And he was right to be pessimistic, as it turned out.

Two months after my visit, the Kenya Wildlife Service arrested Sandy and accused him of poaching; they claimed that he had poached the meat that he fed to his lions, an accusation he vehemently denied. Behind it all lay an apparent land grab by a wealthy local official—land on the Athi-Kapiti Plains is among the most sought-after in Kenya and it will one day be consumed by Nairobi's urban sprawl. Sandy was driven from the conservancy, and two of his lions died during the whole grubby episode.

But all of this lay in the future as Sandy's captive lions roared their disapproval across plains that fell silent and still, as if trying to place the memory.

~

Seven years passed after Meiteranga's first kill. In that time, he killed a second lion, then a third. After he killed his fourth lion, local police arrested him, threw him into prison and beat him. This distressed Meiteranga less than the price the police extracted from him: he was forced to sell two of his best cows to pay the fine of 70,000 Kenyan shillings (close to $1000). This was more than an average Kenyan earned in a year, and quite a sum for a traditional Maasai who largely existed outside the cash economy. The police assured him that worse awaited him should he kill again.

But two days after Meiteranga's release, two of his family's other cows went missing and, perhaps inevitably, Meiteranga again found himself once again pursuing lions. He was certain a lion had eaten the cows and he wanted revenge.

It was 2006 and things had never been this bad: lions were killing Maasai cattle like never before, the Maasai had declared war on the lions of Amboseli, and the act of killing went from an occasional rite of passage to a casual blood sport that would soon empty Maasailand of lions, perhaps forever.

With the sun high overhead, Meiteranga saw two lions—a lioness and an adult male—sleeping under a tree. Eager to claim the prize for himself, he didn't tell the others. He inched closer. With every crack of a twig he froze, hoping that the hum of the African wild would cover his approach; the lions slept on. Meiteranga's heart raced. He could hear the lions breathing. He stared intently as flies buzzed around him. Otherwise, all was still and quiet.

In a single fluid movement he cried out, as is the Maasai way, shouting his family's name for all to hear—'Saitoti!'—as he plunged his spear deep into the side of the startled male, aiming for the heart. Disoriented, the lioness shot off into the bush, but the stricken male, Meiteranga's spear still lodged in his side, ran towards the other Maasai, perhaps a hundred metres away. There they rained spears down upon him and he was quickly killed.

As Meiteranga watched the lion writhe in agony, he felt good, not least because it confirmed him as one of the best and most ruthless lion-killers in a generation: to kill a fifth lion, he knew, would only add to his legend. Only his father—killer of nine—had slain more, and it was, he felt sure, only a matter of time before he challenged that record.

When the lion finally fell still, and with the lioness cast to the wind, Meiteranga took out his knife, removed the mane—a scalping—then cut open the lion's stomach, looking for the vindication he was sure would be his. 'He looked fat!' he later told me. It never occurred to him that he might have been wrong about the fate of the Saitoti family cows.

The lion's stomach was empty.

For possibly the first time in his life as a warrior, Meiteranga didn't know what to do. In an instant, the black-and-white certainties of his Maasai life dissolved, and he sat down in the dust, somehow smaller, his cloak stained with the blood of an innocent lion.

Ordinarily the news of a killing would reach home long before the victorious hunters and, amid an accumulating awareness that a lion was dead and that the party would soon begin, Maasai of all ages would stream in from the surrounding bush to join the final triumphant march of returning heroes. Children would tug at Meiteranga's robes as young women, hoping to catch his eye, clustered in small groups along the way. This time, Meiteranga trudged home in stony silence.

Meiteranga had thrown the first spear and the others knew that they must take their cue from him, an elder statesman among lion-killers. Nobody spoke. Without saying why, Meiteranga made it clear that there would be no celebrations that night. I imagine the other young warriors casting sideways glances at one another, then at Meiteranga, trying to make sense of what was happening. Somewhere along the way, Meiteranga did the unthinkable: he threw his trophies—the mane, the lion's tail, one of its paws—into the bushes.

Never mind that these were powerful symbols of Maasai manhood. Nor did it matter in that moment that a Maasai elder customarily passes such a bounty on to his sons as a symbol of the father's greatness. He would later say that he threw them away to discard any evidence that could tie him to the crime, that he had no desire to return to prison.

But there was more to it than that. 'After four lions,' he told me later, 'I felt that I had fulfilled my greatest wish, which was to get a lion name, and I had that name from killing the first lion. When I killed the others, they didn't add a new name. So there was nothing of importance that I received.'

Such an answer suggests a change of heart, albeit one rooted more in boredom, or perhaps in the desolation of success than any philosophical misgivings about unnecessary bloodshed. Even years later, Meiteranga feigned disinterest when I asked about his refusal to celebrate. It was a quality that came easily to this remote, dignified man who, like many Maasai, was difficult to read and to know. At all times, he maintained a veneer of equilibrium; nothing unsettled him. Nor did Meiteranga boast of his achievements—they were, in his eyes, his due. His actions spoke for themselves, and, anyway, there were plenty of Maasai on hand to sing his praises.

But there was something else—something that he would, reluctantly, come to identify as a nagging awareness of the futility that comes from an unjustifiable killing.

'I felt both angry and sad,' he admitted at last. 'From that moment on, I knew I could not kill another lion. It was a waste.'

Meiteranga is not a man to be pushed and his companions on the day of his last kill knew better than to do so; together they drifted off, muttering darkly, understanding nothing. Alone, Meiteranga passed a troubled night.

His mood worsened the following day. The male lion had not eaten the two missing cows; that he already knew. But nor had any other lion: he found the cows, alive and well. Meiteranga was pleased to find them, and equally so to have escaped prison and a beating—only a handful of Maasai knew what had happened on the hunt and they weren't about to tell anyone. And yet the cows' unharmed presence among Meiteranga's herd was an accusation that went to the core of who he was.

There is no more symbolic act among the Maasai, no act that carries such primeval power, as killing a lion in hand-to-hand combat. For centuries this rite of passage has been a corner-stone of Maasai identity, an act of extreme bravery in which the mantle of the warrior and protector is passed from one generation to the next. As each generation of young Maasai men matures and its members become warriors, this ultimate proof of a warrior's courage becomes part of the coming-of-age story for an entire cohort of Maasai. In these Maasai rites, the lion is at once symbol of a dangerous, hostile world that threatens their very existence, and touchstone by which all Maasai males measure themselves.

'Killing lions is what I grew up with,' Meiteranga told me, no hint of apology in his voice. 'My ambition was to kill more than my father—to kill ten.'

For all of that, Meiteranga knew that killing lions was no casual act, or one to be undertaken lightly: lions were worthy adversaries that deserved respect. 'The Maasai have a very strong relationship with lions,' Eric Ole Kesoi, an experienced Maasai warrior, told me. 'It's part of what makes us Maasai. Yes, it kills your livestock, but there is an unbreakable bond between the Maasai and the lion, and a very strong mutual respect: we respect lions and they respect us.'

That night, after he found his unharmed cows, Meiteranga tossed and turned. He knew that he had hunted lions for the last time. But who was he if not the killer of lions?

Meiteranga Kamunu Saitoti would never be the same again. And what he did next would reverberate throughout the Maasai world.

Like so many African stories, Maasai history began with a journey.

As the Maasai remember it in their telling of their own history, their ancestors inhabited a deep crater or valley far away to the north. Plagued by drought, with the deserts of the north creeping southward, they climbed from the valley, yearning for a greener land and longing for a better life. When they were still within sight of their former home, a bridge collapsed and many died. Those who had already crossed, those who led their people into exile, would become Maasai.

I have seen Maasai herds on the move in times of drought, the cattle as gaunt as their owners with ribcages pressed tight against stretched leathered hides—just as they must have been in the Maasai creation story. They moved, sullen, across dust-bowl plains and eroded gullies where rains had not arrived. Whirlwinds of dust were everywhere, gathering pace across lands picked clean by ill winds, where thin thorns grew and very little else.

I can well imagine those who would become Maasai as they funnelled down through the Rift Valley, conquering and driving into hiding all manner of peoples as they went. I can picture, too, the peoples who stood in their path, watching the horizon as rumours of a strange approaching army swirled around them, their awareness of impending catastrophe as a great cloud of dust stirred in the distance; then a vast army, red and medieval, swarming along the valley floor, spear tips glinting in the hot sun. The peoples who saw them were right to be afraid.

So successful were the Maasai that those who lived in these lands before them disappeared from history. The pre-Maasai past, too, seems to have disappeared: tempting as it may sound, the Maasai are not one of the Lost Tribes of Israel, their ancestors were not members of a lost Roman legion and nor, as far as we know, do they descend from the Ancient Egyptians as some scholars have speculated. By leaving behind their past, the Maasai cast it aside entirely and became something new.

Having reached what is now southern Kenya and northern Tanzania, the Maasai stopped, partly because they liked what they found and partly because formidable enemies—the Kamba to the east, the Barabaig to the south—blocked their path. There the Maasai remained, lords of their loosely defined Maasailand home: a vast sweep of arid savannah in Kenya from what is now Amboseli to the Masai Mara to its north-west, and across the border into Tanzania, from Mount Kilimanjaro to the Serengeti. They were a feared people, raiding sedentary communities beyond Maasailand, and impregnable wherever their home fires burned. Such was the golden age in Maasai memory.

Many Maasai will tell you that the history of their people divides neatly into a glorious past, when the Maasai were truly masters of their land, and a dark modern era. These are the stories that the Maasai tell themselves, and in this they are no different from peoples the world over, peoples whose relationship with the modern world is troubled, whose tales of former greatness mock the subject people they have become. In Africa it is a familiar motif—that of a lost heroic past cruelly cast aside by the coming of the white man. It is, of course, never that simple. But nor is it untrue.

Unusually when it comes to such stories, an account survives of the historical moment when the Maasai turned the corner into despair. In 1883 British explorer Joseph Thomson, at the head of a Royal Geographical Society expedition into the East African interior—the first of its kind in Maasailand—heard rumours of Maasai war parties 2000 strong. The Maasai harried Thomson's caravan and made the Englishman's journey a miserable affair. Even so, there was admiration amid the fear: 'Passing through the forest,' Thomson wrote, 'we soon set our eyes upon the dreaded warriors that had been so long the subject of my waking dreams, and I could not but involuntarily exclaim, "What splendid fellows!"' Later he described a small party of Maasai elders as 'magnificent specimens of their race, considerably over six feet, and with an aristocratic savage dignity that filled me with admiration.'

As I drew near to Maasailand more than a century later, I reread Thomson and was struck by the enduring magnificence of the Maasai profile—the defiant, erect posture, the strange beauty of their red cloaks silhouetted against a land both drained of colour and never more than one missed rainy season away from environmental calamity.

And so it was back then. After Thomson's first passage through Maasailand, everything fell apart, as the Maasai had known it would.

Ancient Maasai prophecies warned that white or red men would come and bring disaster upon the Maasai and their cattle, and that their arrival would be heralded by a comet flashing across the sky. Others foretold that the signifier was an iron snake crossing the land; many have interpreted this to predict the Mombasa–Kampala Railway. Whatever the signs were, on Thomson's return journey a year later everything had changed, and the Maasai assailed him with their tales of woe, blaming him for rains that never came and for the deaths of untold numbers of cattle.

The decade from 1884—the decade that followed Thomson's passage through Maasailand—is known among the Maasai as *enkidaaroto* ('the disaster' or 'when the cattle died'). Settlers soon followed in Thomson's footsteps, and the Maasai who confronted them were weakened by smallpox, their herds devastated by an epidemic of rinderpest and by drought, and their social fabric torn asunder by a civil war between rival Maasai clans. The Maasai population fell by as much as a half, and 90 per cent of Maasai livestock died. Instead of uniting to face a common enemy, Maasai society was deeply divided and in no fit state to defend its land.

Their timing was unfortunate. By the time they emerged from their trials, the Maasai were no longer rulers of their world. In 1901 the commissioner of the East Africa Protectorate described the Maasai as 'the most important and dangerous of the tribes with whom we have to deal in East Africa', adding that he believed 'it will long be necessary to maintain an adequate military force in the districts which they inhabit . . . It would, of course, be unwise to irritate them, and there is always some danger of misunderstanding.'

But colonial caution was tempered with impatience. Two years later, the same British commissioner sniffed: 'As a matter of principle, I cannot admit that wandering tribes have a right to keep other and superior races out of large tracts merely because they have acquired a habit of straggling over far more land than they can utilize.'

It was this latter view that prevailed. In 1904, the Maasai signed a treaty that provided for two Maasai reserves to be established: one in the southern Kajiado region that encompasses Amboseli, and another in the north. The treaty required that the Maasai leave behind—voluntarily, in theory—vast swathes of hitherto Maasai land in the Rift Valley and move to the reserves. But as one British observer would later write, 'finally under heavy pressure the Maasai surrendered, much against their will, to the wishes of the Government ... the whole episode was an eviction and nothing else'.

Worse, the treaty had been 'signed' by an illiterate spiritual leader (*laibon*) called Olonana ole Mbatian. Leaving aside his inability to fully understand the terms of the agreement, his credentials for signing the treaty were flimsy—the acephalous Maasai had no king and there was any number of candidates who could have spoken on their behalf. When the Maasai took the matter to the colonial courts, the East African Court of Appeals upheld the treaty.

In a process repeated across the globe whenever colonial forces conquered and co-opted supposedly less civilised peoples, such treaties became a straitjacket for the Maasai throughout the twentieth century. Less than a decade after the reserves were created, the authorities uprooted tens of thousands more Maasai, and hundreds of thousands of cows, sheep and goats from other formerly Maasai lands, and corralled them into the reserves. More people and more livestock on less land: the pressure from overpopulation and overgrazing intensified and by the 1930s much of Maasailand was at risk of desertification. It is much more so now.

The British, like the Kenyan authorities today, blamed the Maasai for degrading the land and threatening wildlife. To make matters worse, the British seized control over the trade in livestock and sought to limit the size of Maasai herds, a slight that went to the heart of Maasai society.

Not even the reserves belonged to the Maasai. In the 1940s, the British chipped away further at Maasai territory, appropriating land with the best soils and waterholes. In time, from former Maasai territory the British carved a series of national parks and game reserves: Nairobi National Park became the country's first in 1946, followed two years later by the Amboseli National Reserve. Kenya's British rulers at first permitted the Maasai to graze their cattle within the boundaries of Amboseli, but the colonial authorities soon declared much of the reserve to be a livestock-free zone.

They also banned lion hunts. Not for the first time, the Maasai were left mystified by laws that allowed white people— white hunters—to kill wild animals with a gun in the name of sport, but they, the Maasai, were punished if they speared a lion that had killed their livestock. Even so, lions remained plentiful and the Maasai openly hunted them until well into the second half of the twentieth century.

For the October 1954 issue of *National Geographic*, the writer Edgar Monsanto Queeny—part of the Monsanto dynasty, then-chairman of the Monsanto corporation, and self-styled conservationist—and his team *paid* a group of Maasai warriors to go on a lion hunt, and then paid them again when the *National Geographic* team didn't capture the coup de grâce on film; clearly times have changed. Reading the following account, however, one is left in no doubt as to the bravery of the warriors chasing down a lion:

> As we drew near, the lion displayed concern. His tail came up. It twitched. Then the *murran*, out of control, leaped out of the trucks. The lion ran off. He took to a small thicket and, before our cameras could be properly trained, spears were flying. Through my viewfinder I saw the lion rampant, standing on his hindlegs, forelegs raised as if in heraldic emblazonment. He roared, not the earth-shaking roar often heard at night, but one blended with a moan. He turned. Another shower of spears from the left! More roaring moans in quick succession. He fell! And it was over. Not ten seconds had elapsed since the first spear was thrown. Several over-excited *murran* were having fits. One fell to the ground in a coma. Other warriors came out of the thicket proudly displaying their soft iron spears bent, twisted, and bloody. In his agony the lion had rolled over and over, making the spears do double duty by tearing his entrails as he went.

But such accounts would soon become anthropological relics. Increasingly, when the Maasai killed lions, they did so far from the cameras and the risk of punishment.

In 1974, Kenya's government formally declared Amboseli to be a national park. Under the agreement that marked Maasai acceptance of the park's presence in their midst, the Maasai moved onto the neighbouring communal ranches in return for promises relating to park revenues and the provision of water. These promises were, of course, broken, but the park's existence was by then irrevocable. Three years later, Kenya's government banned hunting in all its forms throughout the country.

As one Maasai elder told a lion researcher, Dr Leela Hazzah, in 2005:

> Let's be honest, now. If I take your property away from you without your consent, will you be happy? Now for the fox [government] who forced us to move from the park, have they succeeded in containing these animals in that park? Have the wildlife stopped roaming all over the group ranches and eating our cows? So they could have just left it and just let us live with them like the time before. We have lost our land and our rights to use it.

To this day, the Maasai receive only 4 per cent of park revenues. Although Kenya's government tried to return the park to the Maasai in 2005—a move later overturned by Kenya's High Court—the story of the Amboseli Maasai is ultimately one of dispossession.

~

Leaving behind the Athi-Kapiti Plains, my driver, Peter, and I headed south-east, bound for Amboseli. At Emali, we turned off the Nairobi–Mombasa highway and took a good road through bad country.

As we sped along the ribbon of tarmac that unfurled all the way to Kilimanjaro, a new generation of Maasai men in mud-painted pigtails waved at passing vehicles from the roadside. Few vehicles slowed, but most were oblivious to the makeshift camps where these desultory young men parodied traditional dances with an edge of aggression, an open challenge as much as an invitation.

At Eremito Gate, a queue of safari vehicles strained to enter Amboseli National Park. Old Maasai women with shaven heads and cavernous, empty earlobes milled among the vehicles, offering up carved crocodiles and giraffes; young Maasai men sat in the mid-afternoon shadows.

Here on the cusp of the park, the land was grey. Passing vehicles and even the merest breath of wind stirred up clouds of fine, talcum-like dust that hung in the air, draining colour from the day.

The Maasai stayed outside the gate. We paid our park fees and left them behind.

A corrugated track crossed Amboseli's empty plain. It was a dust bowl, an emptiness. There were no plants to speak of, no wildlife, no Maasai. There were no signs of life.

It was not quite what I expected. There are few more recognisable snapshots of the continent than elephants grazing in green-as-green Amboseli grasslands with snow-capped Mount Kilimanjaro in the background. Those photos were taken here. And, like most snapshots, they substituted an image for an entire story.

Amboseli inhabits the rain shadow of Kilimanjaro and, by annual rainfall alone—as little as 300–350 millimetres per year—it almost qualifies as a desert. Even its name comes from the Maa word *empusal*, for which meanings vary depending on the source: 'salty dust', 'open plain' or 'barren place'. And so it is in Amboseli for much of the year—for many years at a time, in fact.

Back in 1883, Joseph Thomson and his expedition passed close by. 'In spite of the desolate and barren aspect of the country,' he wrote, 'game was to be seen in marvellous abundance . . . The inquiry that naturally rises to one's mind is, how can such enormous numbers of large game live in this extraordinary desert?'

More than a century after Thomson's visit, in 2003, British naturalist and writer George Monbiot described Amboseli as a 'dying savannah': 'The basin which had once displayed one of the world's most spectacular concentrations of wildlife was now, in places, as bleak and dreary as an empty playing field . . . In the mirrored wilderness of grey land and grey sky, even the elephants were dwarfed.' By 1993, Monbiot wrote, parts of the basin were so damaged that Amboseli was closed temporarily after flooding added to the damage caused by over-tourism.

In October 2011, just days before my arrival, Kenya's tourism minister threatened to again close Amboseli, this time for three years, to allow the park time to recover from drought, overexploitation by tourism, and an overpopulation of elephants. Fifteen hundred elephants then inhabited the park, up from 600 a decade earlier.

One suspected the minister of bluffing; Amboseli was too lucrative to close. But the droughts were getting longer and wildlife numbers were radically out of balance with the land—either present in great numbers, or barely present at all. Dr Darcy Ogada, the American conservationist I met in Nairobi, described Amboseli to me as 'a dying ecosystem', and a collapse caused by drought, unsustainable elephant numbers, and an irreversible crash in wildlife populations sparked by mismanagement had come to seem, by the time of my visit, a very real possibility.

These thoughts weighed heavily upon me as we drove deeper into the park until—quite suddenly, close to the park's geographical centre around the Olokenya Swamp—Amboseli became something else altogether. African jacana stepped daintily through green shallows and an African fish eagle circled overhead. Elephants wallowed knee-deep in the water, accompanied by zebra and wildebeest. They were attended by egrets, who feasted on insects stirred up by the hoofs of the striped horses and bearded gnus. On that bright Amboseli day, beneath scudding banks of clouds, southern Kenya's bounty crowded into this narrow strip of swampy earth.

Amboseli's swamps—Enkongo Narok, Longinye, Olokenya—are the oases to the semi-deserts elsewhere in the ecosystem. Most of the rains that fall on Kilimanjaro don't land directly onto Amboseli. Instead, the water filters down into the park via a network of subterranean aquifers, filling swamps and water-holes. In all but the driest years, these small, verdant hollows provide water and food for vast herbivore populations, just as they once served as dry-season waterholes of last resort for the

Maasai. In good years, Amboseli is once again a land of plenty, a breathtaking spectacle and a byword for the abundance of wildlife on the East African plains.

Feast and famine have always been the hallmarks of Amboseli.

But even here, what lay beyond the park's boundaries made its presence felt. A Maasai herder led his cows to water well inside park boundaries; again the pigtails, the aggressive posture, the unspoken desire to provoke. Draped in red and carrying a stick, the herder walked boldly, with no suggestion that his presence was pushing the boundaries of legality—the Maasai may enter the park with their cattle only at times of drought and in ill-defined cases of 'necessity'.

Away to the south, on marginal Maasai lands from where the Maasai cattle came, dust devils spiralled across the plains, gathering momentum from the foothills of Kilimanjaro. The old mountain remained aloof, hiding beneath a corona of cloud, its dark slopes deep in purple shadow.

~

With the sun still high, Peter and I left our lodge inside the park and drove out onto the plains of Amboseli. Almost immediately we came across five lions in a thicket of tall, dry grasses. Safari-goers purred in appreciation, congratulating their drivers on a job well done: 'Would you *look* at that? Look how *big* they are! Aren't they just *amazing*?'

It is easy to make fun of people on safari—their love of khaki, their absence of decorum at hotel buffets, their squeamishness when watching predators bring down prey. Of *course* these animals are big, I wanted to shout. They're *lions*.

Having found lions, few of the visitors lingered. Dr Craig Packer, one of Africa's pre-eminent lion biologists, calculated that most visitors to the Serengeti stop for no longer than ten minutes upon encountering lions before moving on. I found this to be as true here in Amboseli as it is extraordinary; I could, and do, watch lions for hours. But I also knew that many of these people, who took their photos and left the scene with indecent haste, had spent a lifetime saving for these short, concentrated days on safari. Doubtless they would savour these experiences more in hindsight than they seemed to in the moment itself. Certainly those around me would never forget that they had seen a wild lion, and that it *was* just amazing. And who was I to criticise their wonder? I understood their exclamations even if I preferred to enjoy it all in relative silence. Their excitement, which I share, is the very reason that I write about lions.

With the safari vehicles coming and going, all the talk about lions and the threats to their existence felt like arrant nonsense. It was ridiculously easy to stumble upon this supposedly endangered big cat lying in the sand alongside one of the park's busiest thoroughfares. *Look for lions and there shall be lions* seemed to be the lesson of the day, even here in an embattled park like Amboseli. Or perhaps the lesson was this: it is the lion's utter accessibility that makes it so difficult to convince people that the species may be in danger. Unlike the cheetah and the leopard, the lion is a cat of few mysteries, one of very few species in the wild that can afford to be visible and at rest.

These cats, all subadult males, the leonine equivalent of adolescents at around three or four years old, slept off the night's exertions, their bellies bloated. They lay in arrogant repose—one even lay on his back, stomach exposed, four legs splayed like a house cat waiting for someone to tickle his tummy. This was a cat with nothing to fear.

But looks can be deceptive. Between 1990 and 1993, in protest against lost watering holes and broken government promises—a protest as old as conservation in Africa—local Maasai warriors wiped out the entire lion population *inside the park*. In time, more lions arrived from the north-east, but any truce was temporary. From 2001 to 2006, the Maasai killed 130 lions beyond the national park in the broader Amboseli–Tsavo ecosystem, including 42 in 2006 alone; one of those killed in 2006 was the male lion speared by Meiteranga. It could have been even worse: for every lion killed, the Maasai launched 50 unsuccessful hunts.

The Maasai hadn't always killed lions in such numbers. For centuries, since long before there were national parks of any kind, the Maasai lived alongside lions in relative harmony. Young Maasai warriors, as we know, made a point of spearing lions to prove their manhood. But such killings rarely had an impact on overall lion numbers.

It has always been the received wisdom among wildlife experts that the Maasai, like many traditional or indigenous inhabitants of wild lands, were natural conservationists who lived peacefully with lions and other wildlife. But perhaps the Maasai were not conservationists at all. Perhaps, instead, Amboseli's populations of human beings, livestock and wildlife were once in balance, and there simply weren't enough Maasai for lions and people to come frequently into conflict.

That changed sometime during the past two decades, just as Dr Darcy Ogada had told me, and it radically altered the Maasai's relationship with lions. Kenya's population was, in the 1990s, one of the fastest growing on earth and the human pressure upon the lands surrounding Amboseli, a pressure that had been building for a century, reached tipping point. Suddenly there were too many people, too many of their livestock and too many lions all in the same place. By one estimate, in the Amboseli region around 100 lions now lived alongside 35,000 Maasai and close to 2,000,000 head of livestock. In the past, when lions had killed livestock it was explained away as an aberrant form of behaviour by old or injured lions; by the 1990s, the closer proximity of lions, people and livestock made it an increasingly normal part of lion behaviour. This changed everything in Maasailand.

When a lion kills a Maasai cow, it strikes at the very heart of Maasai society. 'To understand lion killings,' Dr Leela Hazzah has written, 'you have to understand that among the Maasai, cattle confer more status and wealth than gold. They are used to buy wives, and the more wives a man has, the wealthier he is. Maasai myth conveys the sense that God bestows cows on his favoured ones, and losing cattle is like losing a child.'

And so it came to pass that during Meiteranga's coming of age, hunting lions—which once marked the arrival of a new generation of Maasai warriors—came to serve an entirely different purpose: the expression of a community's anger. This second reason for killing lions, an act of retaliation rather than ritual, is known as *olkiyoi* in the Maa tongue. As the Maasai began to lose more and more livestock to lions, the men killed increasingly in anger and these kills rapidly eclipsed the Maasai's coming-of-age lion hunts. It was the new normal.

At the same time, assailed by drought, the erosion of traditional values and a growing human population on ever smaller parcels of land, younger Maasai abandoned the old ways in droves. Maasai communal land was fragmenting under the weight of human numbers, and agricultural smallholdings began to block wildlife corridors—a process once unthinkable among the semi-nomadic Maasai, who had always looked down upon those who led settled agricultural lives. Add to this the longstanding resentment of a people who felt themselves excluded from traditional lands and conservation revenues and it is extraordinary that there were any lions left for them to kill. That the Maasai turned their anger on the lions is less surprising than that it took so long for them to do so.

Under a darkening sky, the five subadult lions we were watching by the side of the track stirred, rising every now and then to cast lazy glances towards a herd of skittish zebra and wildebeest, who racked their long brains and circled nervously, trying to understand the source of a threat they could sense but not see. Even with no intention of hunting, these lions—in the tautness of the muscles on their flanks, in the narrowing of their eyes as they fixed their gaze on a straggling calf— possessed what Dr George Schaller, a world authority on lions, described as 'the aura of impending violence'.

Their appetites momentarily sated, unable to muster the energy even for a mock charge to break the tension, the lions lay down to rest, there to remain until the day cooled. Here inside the park, this false paradise, humankind and their vehicles were, for now at least, merely background noise.

~

After killing the innocent male lion, Meiteranga continued to herd his cattle and to lead an otherwise unremarkable Maasai life, feasting on blood and milk and passing endless hours recalling past glories. But the male lion's death had taken its toll on Meiteranga, and he bore little resemblance to the young warrior, driven and proud, who had so confidently killed his first lion. His life felt strangely empty.

While Meiteranga may have undergone an existential transformation after killing an innocent lion, those around him had not. He could, for a time, explain away his reluctance to celebrate the lion's death as a desire to escape further punishment from the authorities. For many Maasai, the Kenyan government is an enemy power and Meiteranga's unwillingness to celebrate his subversive act of killing a lion, and thereby escape punishment, may have gone down well at first. And whatever doubts his fellow *murran* may have had, Meiteranga remained, at least in those early weeks, a hero to his generation—they feted him and called him by his lion name, unaware of the changes that were taking place within him.

But the time came when his decision to cease killing lions could be kept secret no longer. The other Maasai warriors—those who needed courage, those with whom he had hunted in the past—came to him whenever a lion hunt was being planned. He turned them away with excuses. The other *murran* cajoled, joking at first, then more serious as they argued with the man whose passion for killing lions had once been unrivalled.

Meiteranga brushed them off, safe in the knowledge that his record spoke for itself.

'I had proven that I could defend my livestock and that I was brave enough,' he told me later. 'No one could call me a coward.'

But they did. In an effort to turn him, the other young Maasai mocked Meiteranga, questioning his nerve, asking openly whether this was really the same brave warrior who had killed five lions. In time they questioned his courage and even whether he remained worthy of his place among the Saitotis, his legendary lion-killing family. Soon the pressure was constant, the insults calculated to sting him into action: there is no greater insult among Maasai than to call a warrior a coward.

Haunted by the death of the male lion, he held firm. But it hurt.

In the year that followed, driven by an impulse that he barely understood, Meiteranga stood apart from his people, no longer the hero, and suddenly the outsider. He had turned his back on a core Maasai tradition but had nothing to show for it save for the ridicule of his peers. And like so many traditional peoples, he found himself cast into an internal exile between worlds, severed from the past and unsure of his future. It was Meiteranga's dark night of the soul.

Whenever I met Meiteranga in the years that followed, the suspense that came from his unnerving, difficult-to-read intensity was often unbearable. He rarely spoke, only occasionally raised his voice and almost never smiled. There were times when I caught him looking at me from a distance, with an unflinching gaze. He was impossible to read, and his tensile strength was impossible to measure; this was something more than just the famed Maasai air of detachment. His strange inscrutability marked him out from his peers, who laughed often. He lay in wait, coiled and ready, but for what? Meiteranga was not a man to grow old quietly, to let his days as a warrior pass in oblivion.

It could have broken him, this waiting. In the end, what served him well as the killer of lions—his patience, his intensity, his desire to stand apart—was also what saved him.

Perhaps a year after he killed the male lion, Meiteranga began hearing news of an American conservation program, the Lion Guardians, that hired young Maasai warriors. The Kenya Wildlife Service, conservation organisations and tour operators in the Amboseli region had a long tradition of hiring elders, not young warriors, to do their community and conservation work. This sounded different.

At first, Meiteranga kept his distance. Never one to leave himself open to failure, he admitted later that he was initially sceptical: 'When we first heard about the program, we didn't believe it would work. We thought it wouldn't last.' But he also knew that the conservationists wanted to speak with him. More than that, the project stirred something within him.

He wasn't one to follow others' lead. Perhaps he wanted to get in early and lead the way. Whatever the reason, when the Lion Guardians program expanded into the Eselenkay region where Meiteranga lived, he left his livestock with his brothers and went for an interview.

The interview was conducted by the two young American conservationists, Dr Leela Hazzah and Dr Stephanie Dolrenry, whose idea this all was.

'We were new to the area when we first came and then we did Meiteranga's interview,' Dr Dolrenry told me in 2014. 'He came in and he's just so serious. Leela was quite excited—"It's a killer, it's a killer," she whispered. But I was like: "That's great, I'm glad he's here, I'm glad he's interested. This is exactly who we want on our team. But we're going to send him out into the bush tracking these lions that we've been studying!"' Dolrenry was initially unsure of Meiteranga's motives. *I don't want him tracking collared lions yet, not until he's into this*, she thought. *Not until we can read him a little bit more.*

So they sent him out as a trial.

At the time, the Lion Guardians had already spent months looking for lions in order to collar and then track them. The locals assured them that lions were common here in Eselenkay and that they frequently killed Maasai cattle. Yet after count-less hours in the field, they couldn't find a single one. It got so bad they began to wonder whether they should abandon the area and try elsewhere.

Meiteranga had no such difficulties. As soon as he began his trial, he found the lions they'd been looking for. One small group of lions in particular—a male and two females—caught his eye, and when it came time for Dolrenry to collar a lion, Meiteranga was adamant that it must be the lioness. The one missing the tip of her tail.

When the night came to collar the lioness, Dolrenry and Meiteranga set out in near-total darkness, tracking the lions through thick bush. Dolrenry watched Meiteranga for any signs of emotion. Save for the briefest of moments—'When I darted the lioness,' Dolrenry remembered, 'she turned and smacked her sister. I think Meiteranga almost smiled'— Meiteranga remained cold and without expression.

The tranquilliser dart soon did its work and the lioness lay down to sleep. Once they were sure she was unconscious, and that her companions had run off into the bush, they went to work.

When the Lion Guardians collar a lion, the scientists make sure the new Maasai recruits do more than just hold the torch: they do the measurements, they help fit the collar, and they keep watch lest the other lions return.

The collaring completed, Dolrenry asked Meiteranga to name the lion, telling him 'this is your lion now'. Meiteranga thought for a moment, before settling on Nosieki. The name seemed logical enough—it was both the name of the area where the collaring took place, and that of a local bush with pretty red berries and a twisted stem.

As all of this went on, Dolrenry continued scanning Meiteranga for an emotional response, for any hint that what they were doing—laying hands on a lion in deep darkness while other frightened lions circled nearby—was in any way out of the ordinary for him. 'I personally think that they often kill a lion for curiosity. They kill it and they like to play with it, touch it,' Dolrenry told me. 'They see the tracks, they hear them, they see them, the lions eat their cows. But they never get to touch them.'

But as Dolrenry had already sensed, Meiteranga was different.

'He was *very* focused, very serious, and he would do whatever I asked and he would do it really well. The other Guardians, all the other community members I'd done a capture with—they get *excited*. Nothing from him. Just deadpan.'

Meiteranga's calmness under pressure unnerved Dolrenry, to the point that she wondered whether he could be trusted to become a Lion Guardian. *He's too cold*, she remembered thinking. *Even* this *isn't getting to him. This* always *works.* 'I really like these guys,' Dolrenry added. 'They're my buddies and I'm pretty good with them, and I can get even the worst of them to smile. But he was just—nothing.'

Their night's work done, Dolrenry and Meiteranga returned to the vehicle to wait and watch as Nosieki woke. At first unsteady on her feet, she quickly recovered and, none the worse for wear, wandered off into the bushes to rejoin her companions. Dolrenry was also watching Meiteranga, wary of the man who still said barely a word.

'Then, as we start driving away, he gets out his phone and he calls Olubi, another one we'd hired, another notorious lion-killer. He's an older Maasai, one of the originals. He's killed lions here, he's killed lions everywhere. They had already connected because they're old lion-killing buddies. So Meiteranga calls Olubi.'

'Meiteranga!' he said, almost shouting his own name as if about to spear a lion.

'And he just exploded, telling the story, and he was *excited*. That was the moment we got through to him, the moment I knew we could trust him.'

What Dolrenry didn't know at the time was that the lion they collared, Nosieki, was not just any lion. Meiteranga and Nosieki knew each other. Meiteranga had recognised her weeks earlier, and he named her after the place where they'd met: the site of Meiteranga's first hunt as a young *murran*. Nosieki was a cub of the very first lioness Meiteranga had killed, one of the cubs that had escaped into the bush.

It seemed the perfect happy ending: two old foes turned allies in an epic struggle to save the world as they knew it. But there are few happy endings in conservation, and fate was not done with them yet.

~

The sun touched Kilimanjaro's summit but much of the land remained in shadow as Peter and I arrived at Eremito Gate, just before dawn. We were there to meet Mingati, a Maasai guide who was to take us beyond the park into Amboseli's Maasai hinterland. He let us wait, watching without expression from under a tree. When he was ready, he approached the car and climbed on board without saying a word.

At first glance, Mingati failed to conform to the Maasai stereotype. He was neither tall nor slender, nor did he carry himself with that sense of superiority for which the Maasai are famed—he was almost shy. But he did have the muscular frame, that sleek absence of surplus body fat that comes from a life spent on the land.

As with so many traditional Maasai, motley adornments cascaded from his neck—beaded necklaces, coloured strands held together by buttons, even a lanyard of the kind more often used for carrying a USB stick or official pass of some description; Mingati's came from a company called Leo Soccor. Against this vertical assault of colour ran a horizontal tide of belts and bracelets; the latter hugged his wrists and upper arms, and one held in place a green plastic mirror that Mingati consulted from time to time. A plastic mobile-phone case in grandpa-grey was clipped onto his belt. The red and black of the *shuka* wove in and out amid the colours. The overall effect was at once strikingly traditional and that of a Christmas tree hung with the paraphernalia of a magpie-like hoarder.

We shared no common language, Mingati and I, and with brusque hand gestures he directed Peter off the main road and out onto hard-baked plains. There we followed faint vehicle trails among the countless paths worn into the land by centuries of use, and skirted large circular *bomas*, the thorn-brush cattle enclosures within which the Maasai lived in thatched straw huts. Elderly women milled about close to the entrances. Children ran in mock terror as our vehicle approached. Occasionally, two or three *murran*—young Maasai warriors—watched us pass. But the long-horned Maasai cattle that crowd into these *bomas* by night had dispersed for the day, and these small settlements seemed incomplete without them.

All the while, Peter and Mingati got by in the latter's broken Swahili and within no time at all they were firm friends. I have heard it said that the Maasai and the Kikuyu—Kenya's largest tribal group whose lands often border Maasailand—were once sworn enemies and remained suspicious of one another in the forced ethnic proximity of modern Kenya. This was, Peter later told me, untrue. 'The Maasai used to steal our women and raid our farms. But we were never enemies.' I imagined he was joking but when I searched his face there was sign of neither mirth nor sarcasm.

I let the halting, unfamiliar language wash over me, and when Peter translated it was usually because he had heard something he thought would interest me. I had grown to trust Peter's instincts in such matters—he was a thoughtful, knowledgeable man of middle age, genuinely interested in wildlife in a way that is rare among Kenyan guides. One thing he told me was that Mingati had, in his time, killed three lions.

As they spoke, we sped through increasingly harsh country, further from the park and deeper into the Maasai badlands where conflict between lions and people had been rampant. The rains were yet to fall and the land, sandy and littered with thorn, was the tawny colour of marginal lands everywhere in Africa. Here on the desert fringe, wildlife was scarce: a banded mongoose; a squirrel; and a pair of dik-dik, the smallest of the antelope species, cowering beneath low-lying branches, imagining themselves unseen. Livestock stirred up the dust beyond the thorns, pursued by stick-wielding Maasai youth: lithe, draped in blood-red *shukas* and coloured beads, and trailing long ochre-hued pigtails in their wake, they flickered into view like evocations of the African past. One talked on his mobile phone.

The last *boma* marked the otherwise undemarcated boundary of the Eselenkay Group Ranch, one of numerous communal Maasai lands that surround Amboseli National Park. At 748 square kilometres, Eselenkay is almost twice the size of the park. Unlike the other ranches into which the Maasai have crowded since the national park was carved from their midst, Eselenkay was run as a conservancy; although they could graze their herds there, no Maasai lived within its borders. That there was no clamour for this to change owed much to Eselenkay's terrible soils: agriculture and intensive grazing were impossible here, except perhaps after good rains, which were rare. Eselenkay was also home to a luxury tented camp for tourists. Otherwise it was largely empty.

It was in Eselenkay that we were to meet Dr Leela Hazzah, the Egyptian-American conservationist whose work with the Lion Guardians had drawn me to Amboseli in the first place.

Dr Hazzah's camp lay deep within the conservancy, in a small but otherwise unremarkable clearing. The camp felt like a cross between a remote military guard post and an out-of-season university summer camp: spartan army-duty canvas tents strewn with luggage, cardboard boxes and cables; a pit latrine at a polite distance from camp; and a wood-and-canvas tree house that Hazzah and her team called home for most of the year.

Hazzah climbed down from the tree house and we sat in a puddle of meagre shade. Dr Hazzah and her team were warm but wary and, almost immediately, our small talk died; an African mourning dove's call filled the silence. Hazzah made clear that she was meeting me as a favour to my friend Dr Luke Hunter, then president of Panthera, the cat conservation NGO that at the time kept Hazzah's program alive with funding. At times confiding, at other times guarded, Hazzah was as eager to talk as she was to control her message. The conversation laboured as we spoke of people whom we both knew in the small world of carnivore conservation; she studied my face, as if trying to place me in the complicated world in which she moved. I liked her, not least because she was most at ease when talking about lions and the Maasai, a confidence she lost when trying to make sense of the wider web of human relationships.

In time she relaxed a little, and beneath the tree house we drank tea from tin cups as she recalled how, as a child on summer visits to her parents' native Egypt, she would lie awake on rooftops listening in vain for the lions of her father's childhood—lions that had long before vanished from North Africa. She refused to accept that the same thing might happen here, in a lion heartland such as southern Kenya.

From her very earliest days in Maasailand, Hazzah understood that to save lions she must first understand the Maasai. In 2005, when Meiteranga Kamunu Saitoti was still happily killing lions, she moved to Mbirikani Group Ranch, a Maasai communal ranch sandwiched between the Chyulu Hills, Tsavo West National Park and Amboseli's north-eastern boundary.

Although Hazzah did not say so, the Maasai can be a difficult people to know, and it was months before she felt the first signs of acceptance. One afternoon, 'a gathering of about ten *murran* came by my home to drop off a present of fresh milk from one of their mothers to whom I had given a lift earlier that day,' she has written. 'They were setting out to kill a lion that had attacked a cow the evening before. I invited the men in for a cup of traditional *chai* and we talked until late. When dusk approached, the lead *murran* exclaimed: "I guess it is too late to find that damn lion today. Maybe another day . . ."'

Over time, Hazzah's hut became a popular place to pass the time.

'A lot of *murran* would come to my house in the village where I lived,' she told me. 'They came to listen to the radio, or just hang out. The longer I was there, the more they told me about lion hunts. I remember they told me about one lion hunt that they went on in order to prove to the game-scout employers that they should be employed rather than the old game-scouts. They were angry because they were young, fit and could naturally navigate the bush, but all the jobs went to older Maasai.'

It was then that her idea was born.

Most conservationists have one big idea, and Hazzah's was rather ambitious: to take former lion killers and, by co-opting Maasai traditions, turn them into the protectors of lions—or, as she called them, Lion Guardians.

Although the Lion Guardians plan was, back then, no more than a promising idea, a skein of disparate historical strands was coming together: the Maasai were massacring the region's lions, Meiteranga would soon have a change of heart about killing lions, and Hazzah imagined a solution.

'In choosing our first Lion Guardians,' she told me, 'we gave priority to past lion-killers, because they had the respect of their communities.'

Hazzah knew from the beginning that getting someone like Meiteranga on board could change the whole dynamic. As the killer of five lions, he was essential to the program's credibility: if this famed lion-killer could renounce the killing of lions, then the Lion Guardians idea might just work, and the rest would surely follow.

'When we first got him on board, we asked him why,' Hazzah told me that morning in Eselenkay. 'He said that it was because he wanted to be the best at it, just as he was the best at killing lions.' By becoming a Lion Guardian, Meiteranga would once again have a reason to feel proud. It was something he could believe in.

It also gave him and the other Guardians some important new skills.

'Most of those we chose had no formal education. We taught them to read and write, and how to use the radio receivers. And we only selected those who still lived as traditional Maasai.'

Once selected, each Lion Guardian tracked lions across 100 square kilometres of Maasai land and warned herders of the areas to avoid. Around us as we spoke, mobile phones rang often as Lion Guardians in the field called in with news of lion whereabouts and potential points of conflict.

The Guardians also helped to rebuild and reinforce those *bomas* where lions had attacked cattle. They tracked down lost livestock. And with herding increasingly the responsibility of young boys unskilled in the ways of bushcraft, the Guardians brought home lost herders—eighteen in one year alone. Many of them were children.

Key to the success of the Lion Guardian idea, the reason why it worked, was its fidelity to traditional Maasai values: through a clever appropriation of Maasai tradition, the Lion Guardians performed the same function as the warriors who once killed lions. The raison d'être of Maasai warriors—the very reason the Maasai *have* warriors—is to protect their communities. Where the *murran* once fought off lions or led their people into battle against hostile human invaders, the Lion Guardians now helped their people to protect their precious livestock by preventing conflict before it could begin.

In the old times, *murran* gained prestige for their bravery, killing lions with spears—'It is not every young man that can wait for a lion until he is close enough to hunt him with a spear,' one Tanzanian *murran* told lion researcher Dr Laly Lichtenfeld in 2005. Now they won respect by earning salaries, by gaining an education and by holding positions of responsibility. They also proved their bravery by the not-inconsiderable act of staring down a posse of *murran* eager to exact retribution for the lions' cattle-killing.

'Preventing a lion hunt is much more difficult than facing the lion itself,' Meiteranga told me later, when he had had ample experience of both. 'Because if a lion tries to approach and wants to confront you, it's easier because if you stand your ground there is a possibility that you can go back before it reaches you. But dealing with the warriors who are very angry, who are very emotional, who are as bright as you or even more, it is very, *very* difficult, because they can use that emotion, that anger to both harm you and still go after the lion.' In the two years before I met Hazzah, Lion Guardians like Meiteranga had helped to stop almost 100 lion-hunting parties.

Other elements of traditional Maasai culture also came into play, although some owed more to serendipity than careful planning.

'Traditionally when a warrior kills their first lion, they are given a lion name that they will have forever,' Hazzah explained. This was indeed how Kamunu Saitoti had become

Meiteranga. 'Now they track a lion and it becomes theirs. They name the lion.' Just as Meiteranga had with Nosieki.

'Naming has been far more important than we ever imagined,' she continued. 'Now we actually see Maasai communities mourning lions, because each lion now has a name and a story.'

Initial results suggested that Hazzah and her team might be onto something. The first five Lion Guardians began work in January 2007 and, by the time of my first visit in 2011, 40 Guardians patrolled almost 4200 square kilometres across four communal Maasai ranches. The most powerful example of how they fared came from Southern Olgului Ranch, an important wildlife corridor between Amboseli National Park and the Tanzanian border. There, the Maasai killed sixteen lions in the first six months of 2010. Lion Guardians began operating in the area in September of the same year. Not a single lion has since been killed in Southern Olgului on the Lion Guardians' watch.

In the months after my first visit, the successes would continue. In 2011, for the first time since the program began, every adult lioness in the areas patrolled by the Lion Guardians had cubs. New nomadic males also migrated into the area, a sure sign that the lion population was rebounding. By 2013 there were nearly 3.5 lions per 100 square kilometres, up from 1.3 five years earlier. The same year, Lion Guardians successfully tracked down 92 per cent of lost livestock, more than 12,000 cows in total. They also helped reinforce 370 Maasai *bomas*, 93 per cent of which were never again breached by lions. In the seven years to 2013, the Maasai killed five lions in areas where the Lion Guardians were operating. During the same period, in surrounding areas where the program was not in operation, more than 100 lions were poisoned or speared to death.

One thing still bothered me: why, in the end, did Amboseli matter? There were, after all, larger or better protected lion populations in Kenya—in the Masai Mara, Tsavo and Laikipia. Wasn't the tide of growing human populations impossible to hold back? And weren't lions therefore ultimately doomed in human-dominated landscapes, no matter how much goodwill the Lion Guardians generated? It seemed like a whole lot of work, and even more money, to save a small number of lions that would be one day be swept away by the march of human history.

Well, for a start, as Hazzah pointed out, up to three-quarters of Kenya's lions live outside protected areas, and lions such as those in Amboseli are in fact more representative of wild lion populations across Africa than those in the protected Masai Mara. Outside the protected zones, in the places where most of the lions lived, new strategies were needed.

'People have been saying for decades that conservation will only work if it comes from and benefits the communities,' Hazzah told me, with the sun already high over the Lion Guardians' Eselenkay camp where we spoke. 'Well, that's what we're doing. Whatever you call it, whether it's Lion Guardians or something else, we simply have to find a way for carnivores to live alongside people.

'In parks, the animals are generally fine—you can fence them in if you have to. But we don't work in protected areas, and on these communal lands people have to take responsibility for conservation and have a role in the decision-making. Obviously I'm biased, but this is the only way to make it work. There is no other choice.'

She also made clear that my questions came at the problem from the wrong direction.

'Look around Kenya,' she said. 'It's only in Maasailand that lions survive in any numbers.'

Save lions here, her argument ran, *and lions elsewhere might just have a chance.*

Dr Stephanie Dolrenry, then the program's director of carnivore biology, put it in a slightly different way: 'If communities like these can tolerate livestock losses to carnivores, and if we can continue to nurture community engagement so they feel more and more that these lions are "their" lions and they will benefit from having lions, then lions have a future in Kenya.'

And as Dr Luke Hunter, former president of Panthera then Executive Director of the Big Cats Program and the Wildlife Conservation Society, later pointed out to me, if the program were successful its consequences could extend well beyond Kenya.

'Lion Guardians on its own cannot save Africa's lions,' he told me. 'The reality is there is no single silver bullet. But Lion Guardians is one of the most important answers to lion conservation that I've seen. There's no doubt it has turned the tide for the Amboseli lions, and it has potential to do the same for lions in large parts of savannah Africa. The main pillar of Lion Guardians—employing local people to monitor carnivores and help their own community reduce the real or perceived conflict with them—has terrific potential across Africa.'

As I spoke with Hazzah and later thumbed through the reports of the Guardians' achievements, I found myself increasingly excited by what I heard and read. It seemed so devilishly clever: repositioning Maasai values, tapping into Maasai traditions and putting them at the service of lions. It was a rare good-news story amid a litany of conservation woe. The Lion Guardians had me hooked.

~

Later, we piled into the Lion Guardians' battered Land Cruiser. I squeezed into the front seat alongside Hazzah and Philip Briggs, the program's amiable Irish biologist, photographer and later Lion Monitoring Manager, with three Lion Guardians in the back. At a waterhole, we stopped to pick up two more passengers. One was Meiteranga.

Meiteranga.

He came towards the car, walking with feline grace, with the look of a man for whom the world held no fears; one suspected that he looked upon the killing of lions and the arrival of a foreign writer, upon massacre and Maasai rebirth, with the same equanimity.

Here was the Maasai archetype, calm and formidable— the blood-red *shuka* carelessly thrown across his shoulders, the sculpted frame, the beads, the shaven head, the pendulous earlobes. All sinew and taut muscle, neither self-conscious nor afraid, he walked as if he knew the world was watching. It wasn't arrogance. It was the poise that came from utter self-assurance.

I had expected all Maasai to be like him, so closely did he resemble the stereotype. But he was unlike any Maasai I had ever met. In Hazzah's camp alone, one Maasai had buck teeth, wore gumboots and clumped around the camp with a singular absence of authority. Another had the air of a man who would be as at ease in a Nairobi hipster hangout as here on the dusty plains of Maasailand. Yet another was tall but appeared to stoop in apology. Meiteranga was the exception, and he knew it.

As he approached, I watched the other Maasai stand and wait. Only Meiteranga held my gaze. I soon found myself unable to take my eyes off him and there came a moment when I wondered what had just happened, although nothing had; his hooded eyelids gave him a hint of melancholy; here was an enigma.

Meiteranga was a natural leader, a man whose mere presence stopped all conversation. The Maasai waiting by the car parted to let him pass.

He nodded in greeting and we resumed our journey through the bush.

Briggs drove along a barely distinguishable trail through the dust and scrub, following the thin bicycle track left by one Lion Guardian out on patrol earlier in the morning; he had located the lions we were now tracking. Thorny branches grabbed at my arm through the open window; the track was rough, the heat uncomfortable. But this was where we all wanted to be.

We meandered east and then north, looking for lions, talking about them incessantly. At one point, a call came in over the radio to say that 'Qaddafi has been found and killed'.

Hazzah looked troubled and it was some time before she realised that the report referred to the former leader of Libya, not a lion within Lion Guardian territory. 'You do get to become a little obsessed with what you're doing out here,' she admitted.

We passed a desiccated, hollowed-out giraffe carcass, an animal that had been killed by lions a week before. While not unheard of, lions killing giraffes was unusual enough to catch my attention. And it was then that I realised that it was not just the Lion Guardians idea and program that were vital. There was also some really important science going on the background.

So much of what we know about lions comes from Dr George Schaller's seminal 1972 study *The Serengeti Lion*. So thorough was Schaller's extraordinary study that it is still considered by many to be the definitive guide to lion behaviour. But its author cautioned against extrapolating from his results—these were how the lions of the Serengeti behaved: 'It is important to realise that any conclusions about lion behaviour and predator–prey relations refer only to the Serengeti National Park,' Schaller wrote. 'Although my findings may have wider applications, only detailed studies in wider areas can show whether this is so.'

In other words, lions living elsewhere may behave in an entirely different manner. Remarkably little is known about the behaviour and proclivities of lions who inhabit human-dominated lands, but the Lion Guardians team was changing all that.

The most obvious characteristic of the lions studied by Schaller was that they lived in a lion paradise, protected from human pressures by park boundaries and supplied with a regular diet of abundant prey. Beyond the Serengeti, lions have proven by necessity to be a far more adaptable species, willing and able to take advantage of whatever prey is at hand. In the mid-1980s, a pride of lions along Namibia's barren Skeleton Coast learned to hunt seals. In the 1990s, a massive pride of lions in the Savuti region of Botswana took down elephants, as do some prides in Zimbabwe's Hwange National Park to this day. Lions in South Africa's Kruger National Park eat up to 38 different prey species, those in Botswana's Okavango Delta nineteen and those in Tanzania's Lake Manyara National Park just seven. Other reported kills have, according to Schaller, included 'python, catfish, locusts, fruit, and a shirt'.

Although they wouldn't be the first to do so—at least three prides in Tanzania's Selous Game Reserve have beaten them to it—adult lionesses on the community lands of Amboseli have learned to bring down adult giraffes; remarkably, some lionesses have even killed adult giraffes when hunting *alone*. In arid Eselenkay and other non-park territories, this was less about specialising than about lions being forced to take what they could find. Often that meant livestock. But it also came to mean that they learned, Briggs told me, to take down anything. It could be a giraffe, an ostrich or even an aardvark.

The Lion Guardians team was also making other important discoveries that could radically alter our understanding of lions.

Prides of lions are known to inhabit defined territories that they rarely stray beyond, but these home territories are, it seems, more fluid in human-dominated landscapes. Unlike in the relative safety of the Serengeti and elsewhere, most lions outside the protected areas in Kenya are nomadic. Small, easily defendable territories are a luxury for these lions, who must roam further to find food and stay out of harm's way. Within Amboseli National Park, home ranges may be as little as 30 or 40 square kilometres and never more than 250; outside the park, one nomadic male lion tracked by the Lion Guardians ranged across 7000 square kilometres.

Detailed scientific analysis by Stephanie Dolrenry also threw up some startling figures. More cubs—up to 85 per cent—survive to adulthood out here where humans live and hunt lions than they do in the relative protection of the Serengeti; there, the figure is closer to 50 per cent. The logic is simple, if at first glance counterintuitive: the presence of humans keeps at bay spotted hyenas, buffaloes and other species that kill lion cubs. Low lion densities also mean fewer marauding male lions that often kill cubs so as to bring lionesses into oestrus and ensure that it is their own bloodlines, not those of their predecessors, that survive. In the Serengeti, one study estimated, invading male lions are responsible for one-quarter of all cub deaths. Human beings—even the Maasai, it seems—pose less of a threat to lion cubs than do other wild creatures.

~

We left behind the giraffe carcass. Around 14 kilometres from the camp, we stopped and the Lion Guardians among us deployed the high-tech–low-tech methods that are a hall-mark of the project: Leparakuo and Lenkai, two of the Lion Guardians, climbed onto the vehicle's roof with radio receivers, while Meiteranga and Mingati scoured the dust for footprints. The Guardians could discern tracks where I could see only dust, and the receiver gave out a constant beep amid the static. Somewhere in the undergrowth on a broad riverbank, lions watched.

Knowing that wild lions were just metres from where we sat sharpened the senses, and although we couldn't yet see them their presence called the landscape to attention. The signal from the radio receiver was constant; the whispering inside the vehicle fell silent. I stuck my head out the window and stared intently at the bushes.

We waited.

For almost an hour we circled, trying to catch a glimpse of a flicking lion tail, a lion ear, a lion anything. At times the rutted terrain blocked our path, and the Maasai would get down to dig a trail through the riverbed, or to guide Briggs, inching forward, along a safe path. We stopped and listened, and for one moment we become convinced that two lion ears were flicking in the bushes away to our right. Our voices dropped. But more often Eselenkay's ringing hum bore down upon us with the sun.

We slowly became aware that the lions had gone. I wondered whether they had been there at all. It had been a frustrating morning, here in the hardscrabble terrain. As the first clouds of the short rainy season billowed away to the east, we were mindful that the coming rains would soon transform these dusty plains into an impassable quagmire. So we climbed the steep riverbank and turned for home, disappointed.

But a call soon came from the back when one of the Lion Guardians spotted footprints. Suddenly lion pugmarks were everywhere. After an hour in which time dragged its feet, it quickened suddenly and soon we were racing across open country, then diving into thickets, hurtling through the bush, all the while stealing glimpses of lions until they came to rest under a tree.

Briggs did a quick headcount: there were eleven lions. With obvious affection, he explained the history of the group, naming each lion with the familiarity of family members: 'That's Elikan, the female that's snarling at us, and her sister Selenkay.' Eight young lions, from twenty-month-old subadults (Meoshi and Neeki) to four-month-old cubs, rounded out the group. The resident male, shaggy-maned Ndelie, disappeared into the scrub, growling.

This was the largest gathering of lions seen in the area for decades and to find so many lions together outside a national park was almost unprecedented. 'It's incredibly exciting to see such a large group of lions in community areas, especially to keep returning to find that all the cubs are still alive,' Briggs whispered, unable to conceal his excitement as the lions slowly became reconciled to our presence, though watchful in the way of lions in human-dominated lands.

The Maasai in the back seat barely spoke, mindful of another of the team's observations: lions rarely reacted to voices speaking in English or even in Swahili, but ran at the first sound of Maasai voices. Instead, the Maasai watched in awe, and it was difficult to know who to observe: the lions just outside the car window or the Maasai, eyes aglow, who I could see in the rear-view mirror. On one occasion when I looked, I found Meiteranga looking back at me.

Eleven lions were indeed an extraordinary find. Lion prides amid the abundance of the Savuti and the Serengeti can reach close to 40 members. But out here, lions had long understood a simple premise—the bigger the pride, the bigger the target. Here lions had learned to live and forage in much smaller groups and Briggs told me that the lionesses Nempakai and Nolakunte, mother and aunt of Elikan and Selenkay, lived as part of a different group inside Amboseli National Park; in the Serengeti they would have all lived together.

So fragmented were Amboseli's lions that they had, according to Dolrenry, 'a social structure similar to leopards', which are solitary. Lion prides as we traditionally understand them had barely existed in Amboseli until the year before my visit. 'In 2010, that was the first time we started to see multiple generations of females stay together,' Dolrenry told me later. 'We'd never seen that before.' Even so, small bands—a female with a male, a lone lioness with her cubs—remained the norm.

One other thing marked these lions as different: Elikan and Selenkay came close to leonine perfection—they bore no scars or missing ear tips, their teeth were perfect and their tails were intact. Unlike lions elsewhere, these lionesses were not forced to scrap for their meals around a notoriously cantankerous lion dinner table.

From time to time one of the females, Elikan, snarled towards where we sat in our vehicle, just metres away. Her sister

Selenkay watched us, impassive, while one of the younger cubs couldn't resist the urge to play.

'It could just be that these two older cubs will make it,' Briggs whispered after a time. 'But even if just one of the four younger ones survives, that would be a good result. They have to be really careful—in the past three weeks, the pride has killed seven cows and anger among the Maasai is growing.'

But it was the reaction of the Maasai in the car as we finally pulled away from the pride that spoke most eloquently of the success of the Lion Guardians idea: the normally taciturn warriors in the back seat talked all at once, at times shouting in excitement to make themselves heard, recalling the lions' behaviour as if debating the characters in a popular TV series.

These are our lions, they seemed to be saying. *Look how well they're doing.*

~

Barely a week before I arrived, the mood had been very different.

The day had begun like any other at this time of the year. Maasai and conservationist alike emerged from their places of rest close to dawn and looked to the sky, wondering when the clouds that had been building in the east would finally come. Hot winds blew, moving the dry-season dust in eddies and gusts. The merest hint of early-morning humidity clung to shirts and *shukas*, portents of the oppressive October days that precede the rains.

On a normal morning, the Lion Guardians camp is a hive of activity from soon after sunrise, and this day was no different. Early morning is when the Maasai venture out as herders and as Lion Guardians, eager to accomplish tasks before heat slows movement and dulls the senses. In camp, the radio crackles with the static of incoming messages, and mobile phones ring often as Lion Guardians report in.

On this particular morning, both Leela Hazzah and Stephanie Dolrenry were away, and Eric Ole Kesoi, the Maasai community liaison officer of the Lion Guardians program, was at home. When the call came in, it was Philip Briggs who answered. What he heard shook him to his core.

I imagine Briggs sitting there, in that moment when he alone knew something terrible that would change everything. Soon enough, too soon, he took a deep breath, picked up the phone, and called Kesoi. Kesoi, too, knew that the news would send shockwaves through the community. Before they could tell anyone, they had to verify what they had heard. They could not, dared not, tell Meiteranga. Not yet. Not until there was no longer any choice. Reeling, Kesoi made his way to the Lion Guardians camp, where Briggs told him what he knew.

Together with other members of their team, they set off for Osewan, a Maasai area just beyond where the Lion Guardians operate and well beyond the Amboseli heartland. This is Maasai back country, a tough land of subsistence living and droughts that never seem to end. No tourists visit Osewan— why would they?—and its inhabitants receive no benefits from conservation.

In Osewan, the team's worst fears were confirmed.

Stripped of all dignity, the once-beautiful Nosieki, Meiteranga's special lioness, lay crumpled in the dust, her dead cub not far away.

That Nosieki was dead and not sleeping was quickly obvious: curious onlookers crowded around her, with local Maasai taking selfies. Not far away, the carcass of a cow, its ribcage open to the elements, lay rotting in the sun.

In one way, the killing was a return to the pre–Lion Guardians past. A cow had died and the Maasai wanted revenge. An eye for an eye—it is the immutable law of the African wild whenever predators and human beings come into conflict.

Surveying the grim scene, it was easy for Briggs and Kesoi to piece together what had happened. The death was, like most poisonings, a painful one. The lion cub had died instantly but her mother, Nosieki, a powerfully built lioness, had convulsed with muscle spasms not long after watching her cub die. Next, she lost control over her limbs. Confused and desperate, unable to revive her cub, she staggered towards cover. Perhaps she knew that she was dying. Perhaps she had seen other lions die in this way. She may even have heard the mocking laughter of the Maasai herders who had laced the cow carcass with poison, knowing that lions would soon come to feed.

She never reached shelter and died just a few metres from her cub. So strong was the poison that killed the lions that even the flies coating the cow carcass were dead.

Had Nosieki killed the cow? Perhaps, but more likely not; the evidence pointed to hyenas. But such inquests would come too late for Nosieki. In the end, she was just another statistic:

another lion was dead, another lion that Kenya could not afford to lose.

Knowing these lions as well as the Lion Guardians do, they were surprised that Nosieki's male cub had not died alongside his mother and sister. They soon found him nearby in a thorn-brush thicket; he ran off, just as Nosieki had done ten years before him. It was a glimmer of hope, but only a glimmer. Lion Guardian Sitonik gave the fleeing lion a name: Oloishiro, which means 'to be lucky' in the Maa tongue.

With a sense of heavy inevitability, Briggs, Kesoi and the team formally confirmed that the lioness was Nosieki and removed the radio collar that Meiteranga had placed around her neck. They took some samples and photographs and then burned the remains in a hideous bonfire.

As they watched her burn, they knew it was a miracle that Nosieki had made it this far. After Meiteranga had speared Nosieki's mother almost a decade before, the young Nosieki had barely escaped and, frightened and alone, had clung to life. Very few cubs who lose their mothers and are forced to survive on their own are able to do so; the fact that young lions in Amboseli become independent earlier than almost anywhere else may have saved her. Even so, without the protection of a mother, without an adult to teach them the ways of hunting and self-preservation, most lion cubs either starve to death or take to killing livestock, for which they are killed in return in very short order. Having somehow survived, Nosieki led a furtive life, finally beating the odds to begin a lion family of her own but spending most of her time on the fringes of human habitation, dodging the Maasai who would kill her in retaliation for all the ills of their world, and to prove that they were Maasai.

Meiteranga had come to see Nosieki as the bridge between his old life and his new one. It was, at times, a tenuous link. Nosieki had barely escaped Meiteranga's killing spree, and Meiteranga had heard she'd had numerous close shaves in the years that followed. Once he became a Lion Guardian, Meiteranga tried everything he could to keep her out of harm's way. But there was only so much that he could do. Nosieki was, after all, a wild lion. In the manner of wild lions, she wandered away.

And now she was dead.

It wasn't that no one thought to call Meiteranga; no one dared. Again, I imagine those who knew of Nosieki's death shuffling around nervously in the dust, inventing small tasks to postpone the inevitable, knowing what had to be done, hoping that someone else would do it. Perhaps the news reached Nairobi before anyone could bear to tell Meiteranga. Perhaps he still didn't know when Darcy Ogada told me in Nairobi that two lions—these two lions—had been poisoned in Amboseli. But something like this cannot remain secret for long and the team wanted Meiteranga to hear it directly from them.

Before they could call, Meiteranga rang into camp with one of his regular reports. Philip Briggs picked up the phone and told him the news.

~

On our way back to camp after our encounter with Selenkay's pride, we pulled over to the side of the track and Meiteranga and I crouched in the dust to speak, with Kesoi translating. Our conversation, like all our conversations, even years later, was halting and the silences were long. We came from two different worlds, and although Meiteranga had done enough media interviews in his time to know the routine, words with strangers did not come easily to him and he rarely volunteered anything without being asked.

When he did speak, it often had the tone of propaganda, eager as he was to further the Lion Guardians' cause to which he applied himself with the same level of discipline and certainty that he had once brought to his role as a traditional warrior. Every now and then there were fleeting moments of hesitation that I hoped might provide an entry point into a different kind of conversation.

'Before, I killed for traditional reasons,' he told me, 'to prove that I was a *murran*. It gave me great prestige. That was the way it was back then. Now I also get'—and here he paused—'the same . . . satisfaction when I save the lions as when I used to kill them.'

It appeared churlish to point out that satisfaction and prestige were hardly the same thing, and I was left to wonder whether Meiteranga's passion for protecting lions would be quite so strong had he not first fulfilled his Maasai destiny of killing them. Such flickers of uncertainty were, however, rare in Meiteranga. When he continued, it was clear that he believed unreservedly in his role as a Lion Guardian.

'What do the younger *murran* say when you tell them to stop killing lions?' I asked.

'I tell them it's like they're killing our heritage. They say that it's easy for me to say these things because I had my time killing lions. And of course they still have this ambition to kill, to prove that they are *murran*. But they will listen to me because I have killed. I was a traditional *murran* to the hilt. Now they see me on the other side, and how I will do anything to protect my lions.'

Here was the Meiteranga of old as I imagined him in his lion-killing days, bearing himself with a pride that was impossible not to admire. In his new role he could look upon the world once again in black and white, with that steely gaze and with his head held high. But when the conversation turned to the poisoning of Nosieki just a week before, it quickly became clear that he was still grieving, and that the loss was, for him, personal.

'I named Nosieki. She was my lion. I looked after her and protected her. Whenever she left my area, I was waiting for her to return. This time she never did.'

Meiteranga wept upon hearing the news of Nosieki's death—an extraordinary admission for a Maasai warrior—and lost all appetite for food; for a short while, his world fell apart. For the first time since becoming a Lion Guardian, he questioned whether it was all worth it. Perhaps that uncomfortable awareness of life's futility reminded him of those harrowing days after he killed his fifth lion. Sadness and loss overwhelmed him.

Older, half-remembered feelings also stirred, primarily the visceral desire to avenge Nosieki just as he had once avenged his lost cattle.

We will never know how close he came to acting on that desire. All he would say was that he felt as angry as he would have felt if someone had killed a close relative. But deep down he knew that there was no way back—to where should he return? So, with a heavy heart, he went back to work.

He took heart from the reaction of the Maasai elders on the lands where Nosieki was killed.

'In our meeting with the Maasai after the killing,' Kesoi told me, 'the elders of the area cursed it as a cowardly act. "That is not the Maasai way," they said. "Think of the pain the lion died in." Those who killed the lion were embarrassed. Where is the prestige in killing a lion in this way?'

But for Meiteranga, such thin scraps of consolation were drowned by sadness. When he continued, his voice was quieter than before.

'There is a very deep connection between the Guardians and the lions we name. This connection can only be compared to the bond between best friends, or the feelings you have for your best cows.'

I asked if Nosieki could ever be replaced.

Meiteranga thought for a few moments. When he continued, it was difficult to escape the feeling that history might be repeating itself.

'If I can no longer claim Nosieki . . . it would have to be her cub, Oloishiro, that we think might have survived. He would be a very special lion if he makes it.'

But out here, life-and-death struggles rarely come with happy endings and I would soon be reminded of just how rare was the story of Meiteranga and Nosieki. In the months following my visit, Oloishiro disappeared and was never seen again.

Here where we sat in the Eselenkay dust, I well knew that the death of Nosieki was far more than a raw, personal tragedy for Meiteranga and his fellow Lion Guardians, although it was undoubtedly that. Her poisoning struck at the heart of the whole Lion Guardians project. Nosieki had survived Meiteranga's first lion kill. She was present at the transformation of the man who would come to represent the success of the whole program, and she became the most enduring symbol of his redemption. Her story came to stand at the very centre of the narrative surrounding the Lion Guardians. She was the reason why they matter.

We were both silent for a long time. There was damned little to say.

Meiteranga looked off into the distance. While I waited for him to speak, it occurred to me that if this is to be one of the ideas that ultimately helps to save Africa's lions, we may look back upon the death of Nosieki as a defining moment, when those who once killed Africa's lions learned how to mourn them.

Thinking that he had finished speaking, I stood up to leave.

'There used to be so many lions when I was younger,' he said quietly. 'We almost wiped them out.'

~

I returned to Amboseli National Park, and with darkness came the lions.

First Nempakai stepped into the lengthening shadows. She waited, looking around her at the plains with that inscrutable gaze peculiar to her kind. Watchful, alert to possibility, she feared nothing. In time, she resumed her unhurried step, walking out into the failing sunlight with languid purpose. A second lioness, Nolakunte, appeared trailing a cub, then another; two more came gambolling out in spontaneous leaps and playful somersaults, wrestling and swatting then righting themselves to look around in innocent surprise as if caught in an act unworthy of lions. The pride lay down to rest.

The day had been good to these lions: their bellies were full and they had rested, although, being lions, they could not resist the urge to rest some more. Nempakai cradled her head on her paws, lifting it from time to time to regard her cubs with that same feline grace I had seen in Meiteranga. To say that she was at ease understates her position: she reclined, safe in the knowledge that she was the mistress of all she surveyed.

Cars came and soon surrounded the lions, with a whirr of cameras and whispered exclamations of awe; the scene was the East African safari in microcosm. More cars came, others left, but the lions outlasted them all, unmoved by the commotion as the last car hurried away into the dusk, its inhabitants eager to reach the safety of their lodges by nightfall. Out here, the night belonged to the lions.

They had, once again, nothing to fear inside the park. Although they were only 33 kilometres south-east of where we had seen the pride of eleven, Nempakai and Nolakunte—the mother and aunt of Elikan and Selenkay, who ruled the pride beyond the park—were utterly different, and it showed. The lionesses here, even the cubs, regarded the clamour of vehicles in that strange, dispassionate way of lions in protected areas, giving no sign that they had been disturbed. They barely stirred.

But the rainy season was almost upon them. When the rains come, the prey that sustains these lions will move to greener pastures beyond the park's boundaries, and the lions have little choice but to follow.

Blissfully unaware that their prey would soon be gone from here and they would have to leave the safety of the park, Nempakai and Nolakunte stood, stretched and looked towards a distant herd of zebra. Then the lions and their cubs walked towards the park's northern frontier and the lands that lay beyond.

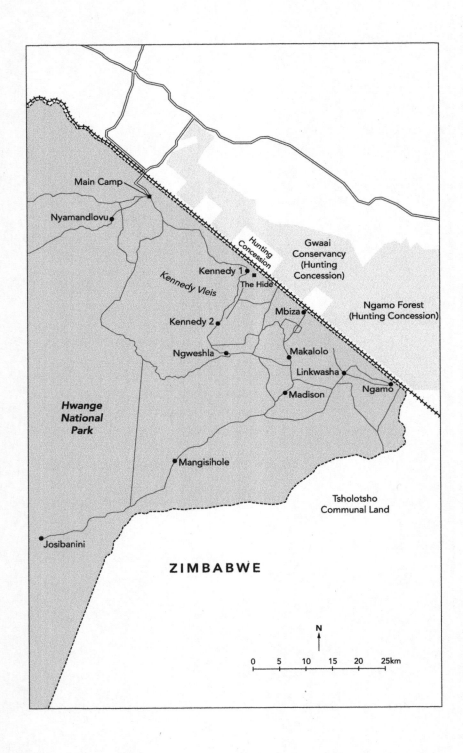

Main Camp

Nyamandlovu

Kennedy Vleis

Kennedy 1
The Hide

Hunting
Concession

Gwaai
Conservancy
(Hunting
Concession)

Ngamo Forest
(Hunting Concession)

Kennedy 2

Mbiza

Ngweshla

Makalolo

Linkwasha

Ngamo

Madison

Hwange
National
Park

Mangisihole

Tsholotsho
Communal Land

Josibanini

ZIMBABWE

N

0   5   10   15   20   25km

2

Cecil the Lion King and Other Stories

Zimbabwe, 2019

Cecil

The Early Years

Cecil was not born into greatness. He took his first steps as a cub in June 2003, not that anyone noticed. Lions come and go here in Zimbabwe's Hwange National Park, and it's impossible to keep track of every new cub. Few if any tourist vehicles make it down to Cecil's place of birth in the south of the park; no one knows who his parents were; and the pride into which he was born deep inside the park was, and remains, a pride of no great consequence. In other words, there was nothing in Cecil's early days to suggest that he would one day become the most famous lion on the planet.

Cecil was four and a half years old when Jane Hunt— Jungle Jane, on-the-ground researcher for the Hwange Lion Research Project since forever—first laid eyes on him down at Mangisihole, on 30 November 2007. A precocious young male,

Cecil wasn't quite the finished article: in a world where dark manes signify leonine strength and power, Cecil's was merely scruffy and trending towards blond, with a hint of ginger. West of where he and his brother, Leander, were first seen, the possibilities for lions stretched far beyond the horizon, extending across the wilderness areas of the park's south, and into Botswana and beyond; the Kavango Zambezi Transfrontier Conservation Area (KAZA), of which Hwange National Park is a part, has its eastern border close to where Cecil was born and continues all the way to southern Angola, thousands of kilometres away to the north-west. East is a different story: no matter where they are in their territory, lions in Mangisihole are rarely more than 10 kilometres from the boundary fence and the Tsholotsho communal lands; lion trails for Cecil and his brother played chicken with people, villages and livestock. If lions stray into Tsholotsho and kill livestock, they stand a good chance of themselves being killed in retaliation.

In September 2008, Cecil and Leander—the Mangisihole Boys, as they became known—were seen again at Josibanini (Jozibanini) south-west of Mangisihole. They were roaming, part of the long process of dispersing from the pride of their birth. Pride males usually expel their own sons, potential rivals, from their kingdoms; the youngsters then roam in search of a territory to call their own. This is the most dangerous time in the life of a young male lion: he must avoid or challenge experienced males who don't take kindly to young pretenders appearing in their realms. Many youngsters head out into the communal lands beyond park boundaries, there to face problems of a different kind. But Cecil and Leander got lucky. By

November 2008, just over a month after they were seen at Josibanini, they had moved in just up the road at Makalolo and Ngweshla; the previous pride males there had wandered outside the park and had been shot by trophy hunters, rich hunters who pay tens of thousands of dollars to hunt a wild lion or other species for 'sport'.

Makalolo may be close to the park boundary, but it's prime lion territory. Both Ngweshla and nearby Ngamo are what Jane Hunt calls 'the pantries of the park' for their abundant prey, good water, and mineral deposits that feed many a prey animal's need for salt. Makalolo forms part of an exclusive safari concession that belongs to Wilderness Safaris, an upmarket safari company with luxury tented camps; vehicle numbers are much lower here than elsewhere in the park. When Cecil and Leander took over this lion paradise, the Hwange Lion Research Project collared Cecil for the first time, and the two boys mated with two Ngweshla lionesses from a band of lionesses known as the Spice Girls. Nearly four months later, the Spice Girls gave birth to six cubs. Cecil and Leander had themselves a pride. Not bad for two young lions with no prior experience of taking over a territory. Cecil was on his way.

The Battle of Ngweshla[1]

There is, of course, a problem with seizing paradise: everybody wants a piece of it. When lions lose it, they want it back. And when they grab it, newly victorious lions just want more. So what happened next was, perhaps, inevitable.

North of Ngweshla, a powerful coalition of four male lions, known as the Askaris, held sway. The undoubted king of the

Askaris was Mpofu, a giant of a lion and one of the great old males of Hwange. Mpofu had lost his three brothers to trophy hunters within the space of just two years. In a masterstroke that ensured that his dynasty would prevail, he teamed up with his three sons—Jericho, Judah, and a lion officially known as Job (to round out the biblical theme) but whom everyone knew as Scaredy Cat. One of the park's most intimidating coalitions, the Askaris had their eye on Ngweshla.

An uneasy truce ensued, long enough for Cecil and Leander's six cubs to be born. But both the Makalolo and Ngweshla coalitions—Cecil and Leander down at Makalolo and Ngweshla, the Askaris up at Kennedy I—would have heard their enemies roaring, marking their territories, waiting for the right moment. They all knew that the battle was coming, and when it came, it unleashed all of the fire and fury that occurs when lions go into battle.

On 22 June 2009, Mpofu led his three sons into battle and together they marched on Ngweshla, hoping to kill Cecil and Leander or, at the very least, banish them from their realm. When battles such as these happen, the pride females run, not so much from fear—although they can be killed by the larger males—but to protect their cubs. It is one of the more unpalatable facts of lion society that incoming males, when they take over a pride, almost always kill all of the pride's cubs. What seems harsh makes perfect genetic sense. Lions live to pass on their genes, to ensure that their bloodlines live on into the future. When the males kill a pride's cubs, females revert quickly to oestrus. To the victor, the spoils: fresh from their victory in battle, the conquering males mate with the females. If they can hold onto the territory for long enough, their offspring should survive into adulthood, and the continuation of the males' bloodlines is assured.

The battle at Ngweshla was an epic, lasting for more than 24 bloody hours. Cecil emerged largely unscathed; for the most part, Jericho, Judah and Job watched from the sidelines. But Mpofu and Leander shed enough blood for all of them. When the fighting ended, Leander could barely walk—'He was limping on three legs,' Jane Hunt remembers. He died a week later. Mpofu, too, was badly injured, and was never the same again. Although Mpofu hung on for a few more months, his body was broken. The park's authorities euthanised him on 2 November, at the grand old age of fourteen. But before Leander and Mpofu succumbed, Cecil did the maths and realised that, alone, he was no match for the three remaining Askaris. He fled.

As victors, Jericho, Judah and Job did what their genes demanded of them: they killed the six cubs sired by Cecil and Leander at Makalolo and mated with the lionesses who had mated with Cecil and his brother. Cecil was cast out into the wilderness.

The King of Linkwasha

That could have been the end of Cecil's story. Condemned by the loss of Leander to a solitary life, he could have disappeared into the south of the park to lick his wounds. A lone male can find it difficult to protect a territory that he already holds; seizing a new territory is even harder. Life as a lone nomad is the lot of many male lions, surviving but unfulfilled, and never able to pass on their genes to the next generation.

Cecil was destined for something different.

By October 2009, just four months after the Battle of Ngweshla, Cecil turned up at Linkwasha. With a name that, according to one guide, means 'pathways', a reference to the many trails made by elephants in these parts, Linkwasha is due east of Makalolo, just beyond the outermost reaches of the Ngweshla pride's range and, like Makalolo, a part of the Wilderness Safaris concession. If there had been a pride male at Linkwasha when Cecil arrived, he was gone by the time Cecil was seen there in October.

Linkwasha, also known as Backpans, was a far more modest territory than the one Cecil had sought with Leander. In contrast, the Askaris' prized pride lands now extended from Ngweshla in the west and south to Makalolo in the east and to almost as far north as Kennedy I. But Cecil's control of Linkwasha would be the making of him.

It was at Linkwasha that he first became known to the guides from the Wilderness camps, and to many tourists. He had grown into a handsome lion, his mane dark and glorious, and he became a favourite of many guides and safari visitors alike. Appear in enough Instagram posts and Facebook time-lines and you're well on your way to becoming something of a minor lion celebrity.

This newfound fame and popularity took some getting used to for Cecil, this lion from the Hwange backwaters. 'Cecil was a fantastic lion, but he could be grumpy at times,' said one expert who knew Cecil well. When animals come in from the Hwange back country, he said, some safari guides drive too close to the lions in order to help their clients get the best photos. Some of the lions 'do grow up around the safari camps', and are more used to vehicles, 'but it's still unnecessary. So when you get some new boys come in and they're not used to that treatment, then you get some excitement. In the end, Cecil grew used to it.'

Those who lived within Cecil's territory also needed some time to adapt. On one occasion, a camp attendant at Linkwasha was walking between the rooms early one morning when Cecil loomed in front of him. Instead of running, the camp attendant was so shocked that he fell to the ground unconscious. Whatever had been Cecil's intention, this unexpected turn of events left him nonplussed and he wandered off. It happened often. 'When Cecil was around,' said Jane Hunt, 'there were many people who had scary wake-up calls, I can tell you. He would often grunt at the guides when they were walking and waking people up.'

But everyone learned to live together soon enough. Cecil flourished and his fame grew. 'Where he became famous in terms of safari guides and tourists was at Linkwasha,' said Brent Stapelkamp, a lion researcher in Hwange from 2006 to 2016. 'Cecil had a big pride, up to 25 lions. They were killing elephants deep in the park, no disturbance, and vehicles that gave him nothing but the sound of Americans. I've seen him walk underneath the tiered seats—they had the white Land Rovers in those days and the final seat was sitting right out the back. I saw him walk underneath the bums of three Americans. He was that relaxed around vehicles.'

Hold on to a territory for long enough, though, and other lions take note. No doubt there were skirmishes as young male lions passed through Cecil's range, testing his strength and resolve. With each victory, his position became more secure. As a lone pride male, he remained vulnerable to larger coalitions and, for a time, the Askaris remained in power just up the road at Ngweshla. But like any territorial male, he patrolled the boundaries of his small kingdom, scent-marking the perimeter; with each month that passed, his sign must have grown in power and significance.

To stay in power, a male lion must do more than ward off challengers. Females, too, live to pass on their bloodlines, to have cubs with strong and desirable males, and to see their cubs grow to adulthood. Thankfully for the females of the Linkwasha pride, Cecil was rather good at protecting the cubs, which was ultimately the most important of his responsibilities. In the three years that Cecil held Linkwasha from late 2009, he sired no fewer than eighteen cubs. One of these was Xanda, a strapping male cut from the same cloth as his father. Of the eighteen cubs, six females were still alive in early 2020 as members of the Linkwasha and Ngamo prides. In them, Cecil's bloodline lives on.

For all of Cecil's success at breeding, a shadow existed—and still exists—over all the lions of Linkwasha. Along the northeastern boundary of what was Cecil's territory, a railway line marks the official boundary between Hwange National Park and the forests that form part of a trophy-hunting concession. On one side of the railway line, lions were protected. On the other, lions were fair game. Except when a train passed by—in addition to the twice-daily passenger service between Bulawayo and Victoria Falls, freight and other smaller trains use this section of track—there was nothing to stop Cecil and the Linkwasha females from crossing into the hunting lands. If you stand on the track and look from one side to the other, nothing distinguishes the two territories. The habitat—acacia woodlands; poor, sandy Kalahari soils; small clearings with waterholes—is the same and there is prey on both sides. And unlike those places where villages and farms press close to the park boundary and the dangers to lions are obvious, in the hunting concessions lions will see none of the triggers—no buildings, no cleared land, no people, no dogs—that have taught them to be careful.

But for now, at least, Cecil avoided trouble and remained the king of his world.

Exile

Just up the road in Ngweshla, life was not going smoothly for the Askaris. In May 2010, just shy of a year after the Battle of Ngweshla, Russian hunters shot Job (Scaredy Cat) just outside the park. His rather disparaging nickname was earned not because he didn't like a scrap—Job showed no sign of being afraid of other lions and was, by all accounts, always up for a fight. He was known as Scaredy Cat because he was wary of humans and their vehicles. Brent Stapelkamp remembers how he was rarely able to capture Job on film—upon seeing Brent and his kind coming, Scaredy Cat would slink off into the bushes. In the end his fear was justified, but it couldn't save him from the hunters' guns.

Six months later, his brother Judah was also shot just across the railway line. Unlike the killing of Job, the hunt that killed

Judah was illegal, although no one was punished due to insufficient evidence.

Both Job and Judah should have been safe. When you stand at Ngweshla, the world of villages and hunting can seem like a very long way away. Vehicles come and go, but they're all tourist or park vehicles, and the straight-line distance to the railway line, the park boundary, is almost 20 kilometres. More than that, between Ngweshla and the hunting concessions lies a string of pride territories occupied by other males. These should act as buffer zones that keep lions from Ngweshla and deeper within the park from straying into hunting lands.

Part of the explanation of why Job and Judah had strayed so far from Ngweshla stems from their own greed. A male lion is ever-restless; his is an endless quest for expansion, for new females with whom to mate, and for food. Hunters are known to use baits, drawing lions to the carcasses of dead elephants and other animals, sometimes using the calls of animals in distress just in case the lions don't get the message. 'They know that whole area,' said Jane Hunt. 'It's right on the edge of where they go. There was free food. They're lazy cats. But you've also got them protecting their territory. And that means going up there and pushing the other folks that are up there out. There are also females up there. These animals with these huge pride ranges are always going to get themselves into trouble, because somewhere along the line, they're going to go into badlands. And by lion standards, it's not far.'

The Hwange Lion Research Project has a further explanation, and it's a profoundly disturbing finding on a continent where human communities crowd up against the boundary of many national parks. By analysing data gathered over more than a decade, Dr Andrew Loveridge, co-founder of the project, found that the killing of territorial male lions left what he called 'territorial vacuums'. In other words, as hunters killed male lions, the former territories of those lions became vacant or, rather, up for grabs. This, in turn, sucked park males from the relative safety of the park's interior out into the boundary areas. 'One home range, around Hwange Main Camp, was successfully filled four times during the study as its occupants were, one after another, removed by sport hunters,' Loveridge wrote. And wherever this happened, this left the females vulnerable to attack and their cubs at grave risk of infanticide.

The implications are terrifying: studies by Loveridge and his colleagues found that any lion with a territory whose centre or core was 20 kilometres from the park boundary was in danger of dying at the hands of humans. Push this out from the centre of a lion's territory to its outermost reaches, and, as one lion expert said, any lion 'within 40 kilometres of the boundary goes out'. What this means is that just 38.7 per cent of Hwange National Park is capable of fully protecting its lions. In two-thirds of this park, one of southern Africa's largest and most important, lions cannot be considered safe.

None of these reasons, of course, mattered to Jericho. Poor old Jericho. A year and a half after they had emerged triumphant in the Battle of Ngweshla, his Askaris had gone from four members to just one. He had seen his father fight and die as a result, and he was with both Job and Judah when they were shot. Now he was alone. That he was able to hold on to one of Hwange's most prized lion territories for nearly eighteen more months was remarkable. During that time he enjoyed the bounty of Ngweshla in the manner of a king ruling over a land of plenty. 'His nickname was "Hoover",' Jane Hunt remembered, 'because he was always the last on a carcass and could barely walk. He used to hoover it all up, old Jericho.'

But it couldn't last, and no one was surprised when two lions known as Bush and Bhubesi appeared on the scene and forced Jericho from Ngweshla in March–April 2012. Jericho fled south and took to killing cattle down on the boundary. A month after he was kicked out of Ngweshla, Jericho was caught in a wire snare that cut deep into his neck. Brent Stapelkamp darted him to remove the snare but was forced to remove the tracking collar to allow the wound to heal.

While all of this was happening, Bush and Bhubesi were harassing Cecil down at Linkwasha as well. They could have kicked him out but they were busy at Ngweshla, so they just let him know that they were around, occasionally mating with his females and generally making his life miserable. By November 2012, Cecil's hold on Linkwasha was waning, and by June 2013, he was on the run for the second time, headed south. Not long after he left Linkwasha, the signal from his collar died.

Cecil and Jericho, two minor giants of recent lion history, had been cast into purgatory.

It is impossible to know what goes on in the mind of a lion. We can only speculate, basing our predictions on long-running studies of behaviour, extrapolating from the past experiences and decisions made by lions under similar circumstances. The first instincts of Cecil and Jericho would have been simply to survive as best they could, taking down any prey they found, defending themselves when under attack and avoiding battles they knew they could not win. But deep down, I suspect—it is not scientifically proven, but I believe it to be true—every male lion who loses a pride and a territory, and even one who has never ruled a pride, still, until his dying day, yearns for power so that he can sire offspring that continue his legacy long after he himself has died. It is the unrequited need at the heart of many male lion stories: some never rule, and many are forced from power never to return.

That should have been the fate of Cecil and Jericho, these two old enemies. Cecil was around nine years old, Jericho closer to eight. Although still in their prime as males, they were close to the age where an ousted male will often spend the rest of his life roaming without a territory of his own, no longer capable of matching the younger coalitions. But one of the great things about lions is that you never quite know what they're going to do next.

Kings of Ngweshla

Jericho was, as they say in these parts, 'in the wind' for nearly eighteen months. Cecil's period of exile lasted for around six months. No one knows where they went—neither had a functioning collar—and people began to forget about them as other lions took centre stage. They couldn't possibly be together. These were lifetime rivals, after all: Cecil's brother had been responsible for the death of Jericho's father, and Jericho's father had killed Cecil's brother.

And then, all of a sudden, in September 2013, a message reached Brent Stapelkamp over the radio, in tones of incredulity: 'Brent, you won't believe it. We've found Jericho and Cecil and they're rubbing heads.' Males will usually team up with brothers or cousins to form a coalition. Occasionally, unrelated males will join together. But for two blood enemies to turn around and become partners is almost unheard of. Years later, Stapelkamp still shook his head in disbelief when retelling the story; as he put it, 'They had met somewhere in the hinterland and worked out their shit.'

Calvet, an experienced local guide from the Somalisa Camp, put it differently when he imagined the conversation between these two old foes: 'Cecil says to Jericho, "You don't have a territory. I don't have a territory. Let's form a coalition, my friend, and let's go to Ngweshla."'

Even Jane Hunt, who seems incapable of surprise, found the whole episode to be 'bizarre'. Her only explanation is that making peace not war somehow suited Jericho. 'It was just very weird that they joined together after being archenemies. I think Jericho's father, Mpofu, still rolls in his grave at his beach-boy son. Jericho was a very placid lion. He was always a lover not a fighter.'

Whatever the reason, Cecil and Jericho marched on Ngweshla and drove Bush and Bhubesi away. They were both big lions—Cecil with his dark, fulsome mane, Jericho with his showy blond head of hair—and they were once again the kings of their world. The females of Bush and Bhubesi fled but they hadn't gone far. The guide Calvet takes up the story: 'Cecil and Jericho—now, there they are, standing at Ngweshla without any girls, so they started advertising. Then three girls from the Spice Girls came, they saw them, and they were *so* happy to see these big guys. They stayed with Cecil and Jericho the rest of 2013 and they mated with them in 2014.'

The Spice Girls gave birth to cubs at the tail end of 2014, and Cecil and Jericho were well on their way to becoming Hwange's biggest lion celebrities. Everyone—from the lion project to the safari guides and their guests—knew Cecil and Jericho and knew their history, their unusual stories of survival, all the drama and suspense, triumph and heartbreak. Jericho in particular had been present at the death of his father and two brothers, and had been snared for straying into community areas. Both had won and lost territories and had sired cubs that still stalked the plains of Hwange.

Cecil's individual reputation, too, was growing. He was easy to approach and was relaxed around vehicles. Safari guides loved him because he had returned to Ngweshla after everyone had written him off. And tourists adored him because everyone on safari wants to see a male lion, and he was a big, beautiful lion, with a big, beautiful mane. With his media-friendly name and social-media-savvy admirers, Cecil was already becoming one of the first lion superstars of the digital media age.

Somewhere over in Minnesota, a dentist named Walter Palmer was watching.

A Pact with the Devil

I must be honest: why anyone would want to shoot a lion, or any other living creature for that matter, for sport is beyond me. The whole idea that an animal mounted on a wall could be somehow more beautiful than that same animal alive and roaming the plains of Africa is something that I will never understand. For all its perceived romance, for all its testosterone-fuelled bravado, the shooting of a lion in the 21st century involves very little bravery, or, dare I say it, sport. A century ago, perhaps, many hunters tracked animals through the bush, often on foot, and they suffered the consequences if they failed to kill the animal with a clean shot. The very notion of the 'Big Five'—lion, leopard, buffalo, elephant and rhinoceros—came from hunters nominating these five species as the most dangerous animals to hunt, those that could kill a hunter if the first shot wasn't successful. Modern-day hunting involves no such risk.

Anyone who has seen a lion in the wild, especially a lion as accustomed to vehicles as Cecil, would know that you could drive right up to him with no danger to yourself whatsoever. These days, modern hunting operators set up a bait, watch from a hide and, in some cases, tell their rich clients when to shoot. There is no bravery involved, no skill other than the ability to shoot straight at a stationary target. The lions that they shoot are not lions unaccustomed to people or to vehicles—there are few resident lions in the hunting concessions and all come from inside the national park. As the Hwange Lion Research Project has found, 82 per cent of all lions hunted in the concessions bordering Hwange National Park are killed within a kilometre of the park boundary.

But when it comes to hunting, conservationists have learned to put their personal feelings to one side, and I shall try to do likewise. That's because, as Andrew Loveridge has written, hunting is 'considered a necessary part of wildlife conservation and management and has been a driving force in conservation since the early 20th century'.

Pre-eminent lion biologist Craig Packer has estimated that '80 percent of the lions left in the world are in the hunters' hands'. To put it another way, hunting concessions 'protect' many more lions than do national parks in Africa. Much of wild Africa, much lion territory, is simply unsuitable for photo safaris or 'photo-tourism'. That may be because it is inaccessible to ordinary safari vehicles, or because the combination of vegetation and landscape score a negative when it comes to the kind of scenic beauty that safari tourists expect from their once-in-a-lifetime African experiences. Or it may be because the continent's governments simply don't have the resources to care for these vast tracts of wilderness—to forge trails, develop the requisite infrastructure, and put in place the necessary anti-poaching measures to protect the land and its wildlife. In 2012, for example, the hunting operators of Savé Valley Conservancy in Zimbabwe were estimated to spend US$546,000 every year on anti-poaching measures and, in 2019, they employed around 200 anti-poaching game scouts. Government investment in the conservancy is minimal. In South Africa, trophy hunting has been credited with playing a critical role in saving species such as the white rhino, bontebok, black wildebeest and Cape mountain zebra from extinction, because the conversion of livestock ranches to hunting concessions allowed these species to flourish.

Another argument in favour of trophy hunting is that it puts a value on wildlife for the local people. Some hunters pay US$50,000 for a permit to hunt a single lion; in Botswana a few years ago, the figure may have been as high as US$140,000. In January 2014, the Namibian government controversially auctioned off a permit to kill a black rhinoceros. Black rhinos are critically endangered, but this particular male was past breeding age, he was considered a danger to other rhinos, and the proceeds were to go to rhino conservation projects in the country. With a winning bid of US$350,000, a hunter from Texas named Corey Knowlton took the permit, claiming he was doing far more for rhino conservation than his critics who flooded social media with their anger. To earn an equivalent sum from photo-safari tourism would require dozens, perhaps hundreds more visitors, increasing the tourism footprint. Apart from the impact upon the poor lion or rhino, hunting can, therefore, leave a far smaller imprint upon the land than your average safari.

Many hunting concessions surround national parks, serving as buffers between officially protected reserves and community areas with growing human populations. One study found that trophy hunting was responsible for the conversion of 27,000 square kilometres of livestock ranches to wildlife-rich hunting concessions in Zimbabwe. Were the reverse to happen, were hunting concessions to disappear, it is a pipe dream to imagine that these lands would be incorporated into national parks. Almost universally, most of the land would be converted to agriculture or cleared for livestock. Much lion habitat—60,000 square kilometres across Africa as a starting point—would be lost in the process.

'Be careful what you wish for' seems to be the argument. 'I say to people, "Don't be unreasonable",' said one lion expert about the situation around Hwange. 'If you take hunting out of that area at the moment, what is land-use going to be? It'll be worse for wildlife if they go back to cattle.' Conservationists don't like hunting, but as one lion researcher told me, 'It's an evil in some ways. But sometimes it's the lesser evil.'

All conservationists and, it must be said, reputable hunting operators agree that hunting can only work as a conservation tool under quite specific conditions. Hunting quotas—the number of lions and other trophy animals that may be legally shot in a particular area—must be based on good science, on reliable lion population numbers gleaned from careful, long-term studies of local populations. It is also critical that only male lions can be shot, and that those males must be at least seven (preferably eight) years old. The latter age requirement assumes that, by this age, male lions have had an opportunity to sire cubs and bring at least one generation through to adulthood. Wherever male lions younger than seven have been routinely shot as trophies, the local lion population has fallen at a rate potentially dangerous to the survival of that population. Systems for age and quota verification, as well as transparency in the process of granting and renewing hunting leases, are also critical.

So much for the theory. Across Africa, adherence to these norms is the exception rather than the rule. The hunting industry in Tanzania, home to an estimated half of all lions left in Africa, has, for the most part, fallen well short of these standards. Elsewhere the story is mixed. Niassa National Reserve in Mozambique is considered a success story. Namibia, too, is held up as a place where sustainable hunting practices contribute to conservation. When it comes to Zimbabwe, a country not known for its good governance, the situation is better than you might expect, although far from perfect. Despite excesses in the past, the hunting industry and pro-hunting government work closely with conservationists, the national parks and the Hwange Lion Research Project to ensure that the trophy hunting of lions is sustainable. Of course, it doesn't always work out this way. Even when you exclude the frequency with which illegal hunts occur, it remains highly questionable whether the trophy hunting that takes place around Hwange National Park falls within the standards necessary to maintain a hunting industry that aids conservation, or at least doesn't harm it.

All of this makes Hwange a good place to judge whether conservation's pact with this particular devil—the mantra that you have to shoot some lions in order to save the broader population—does indeed save lions. I understand the arguments. I have my doubts.

The King is Dead

I wasn't there when Cecil was shot by Walter Palmer. But this is what I believe happened.[2]

We can only speculate whether Walter Palmer specifically targeted Cecil because of Cecil's social media fame as a large, well-known and charismatic lion. It would make sense. According to Brent Stapelkamp, Palmer already held the trophy record for the 'World's Biggest White Rhino' and the 'World's Biggest Polar Bear'. It does not take any great leap of logic to imagine Palmer getting on the phone and telling his hunting operator: 'I want that lion.'

Whether or not it was Cecil's fame that drove Palmer in his choice of lion, there appears little doubt that Cecil was very much the target. How else to explain what happened on the night of 1 July 2015? Within the privately owned Gwaai Conservancy area, trackers working for Theo and Zane Bronkhorst laid out an elephant carcass just 700 metres from the national park boundary. By 9 p.m., Palmer was in the hide with Zane. Although Theo would later take the rap for organising the illegal hunt and was present in the hunting camp, some reports suggest that he was covering for his son, Zane, who was the professional hunter overseeing the hunt and was in the hide with Palmer.

Although the Zimbabwean government and the Bronkhorst family lawyers would later claim that the hunt was entirely legal, it simply wasn't. The owner of Antoinette Farm where Cecil was shot, Honest Ndlovu, was not granted any lions on quota in 2015; he was being punished for having shot, or allowing to be shot on his hunting concession, underage lions the year before; of the five lions shot in the whole Gwaai Conservancy in 2014, only one was found to be six years of age or older. As Brent Stapelkamp tells it, 'It was illegal primarily because there was no quota . . . No paperwork was filed—no pre-hunt documents, no post-hunt documents. The collar was destroyed. We never got it back. For a crossbow hunt in Zimbabwe, you need a national parks ranger with you. They didn't have one with them. For a lion hunt you need a parks ranger with you. There was none. It was illegal for a whole suite of reasons.'

No one knows whether the Bronkhorsts told Palmer that they didn't have a permit; in any case, Palmer has history in this regard. In 2006, Palmer shot a black bear in the US state of Wisconsin, in a county for which he had no permit to hunt.

He allegedly tried to pay nearly US$20,000 to his guides to lie on his behalf, and he was later convicted of the felony of making false statements and shooting an animal illegally.

Whatever he knew in relation to his Zimbabwe hunt, there he was, waiting in the hide for a lion to come to feed. Palmer and Bronkhorst were about 40 metres from the bait. Their scouts told them that lions were on their way. The anticipation was building and, right on cue, Jericho—old Hoover himself, never one to pass up a free meal—walked into the clearing and started to feed on the carcass. Jericho was a magnificent creature. Yes, his mane was blond. But, said Stapelkamp, 'unless you knew that something bigger was coming, there is no way you wouldn't shoot Jericho. For a full hour they watched him feed.' Palmer was waiting for Cecil.

Soon enough, Cecil arrived. He and Jericho may have been new best friends, but lions don't like to share food—teeth were bared and much snarling ensued, until Jericho, the lover not the fighter, yielded to Cecil. Cecil began to feed. With a spotlight trained on Cecil, Palmer raised his crossbow and fired.

What normally happens in these situations is that when a client shoots, the professional hunter in attendance—in this case, Zane or Theo Bronkhorst—is ready: if the client's shot doesn't kill the animal stone dead and the animal runs, the hunter is required to follow up with the killer shot to 'anchor' the lion; no one wants a wounded animal disappearing into the bush. But in this case, in Brent Stapelkamp's telling of the story, Bronkhorst was under strict instructions *not* to follow up. This was a crossbow hunt, after all, and if Palmer was to hold the record for shooting the World's Biggest Lion, he couldn't have someone else deliver the coup de grâce with a rifle.

Cecil ran off, badly wounded but alive.

Let him bleed, you can hear Zane or Theo Bronkhorst telling Palmer, patting him on the back. *Let's go and have a whisky in camp. Give him time to feel it, so he's bled out a bit. We'll get him in the morning.* So they went back to camp, had dinner, toasted their hunt and went to bed.

Cecil hadn't gone far. He couldn't walk more than 100, perhaps 200 metres from where he had been shot. He had an arrow in his side. He must have spent the night in agony.

When Bronkhorst and Walter Palmer returned the next morning—not at 5 a.m. but around nine in the morning—they followed the trail of blood and drove right up to Cecil. And so, very bravely, eleven hours after first shooting Cecil, Palmer shot and killed Cecil with a second arrow, from the safety of his vehicle.

You can imagine Bronkhorst and Palmer getting out of the vehicle. Mission accomplished. High-fives all round. They walked over to the dead lion, and for the first time they saw the radio collar that the Hwange Lion Research Project had placed around Cecil's neck on 28 October 2013. There is little doubt that the Bronkhorsts would have known that Cecil was collared. What appears likely is that neither of them ever told Palmer. And given that Palmer had only seen Cecil from a distance of 40 metres, we must give him the benefit of the doubt that he hadn't seen the collar the night before; in a dense mane like Cecil's, collars can indeed be difficult to see, especially in the dark.

Whatever. According to Theo Bronkhorst's statement to the professional hunters' association, when Palmer saw the collar for the first time, he freaked out. He panicked and ran. He'd just killed a collared lion. 'What the hell is that?' he demanded. Suddenly afraid, he wanted to get the hell out of Zimbabwe. More adept at dealing with wounded animals than clients doing a runner, Bronkhorst hung the collar on the branch of a nearby tree and took off after Palmer. Having consoled his client, having told him that everything was fine—boy, did he get that wrong!—Theo later told police that he returned to the scene of the crime. The collar was gone. Now if you believe that . . .

Then things got weird. Seven hours after Cecil was shot and finally killed, at 4 p.m. on 2 July, the collar started to move. We know this because Cecil's collar was still sending GPS data. Not only did the collar move but, according to one account, it moved like a lion. It walked up the road. It went to drink. It got back on the road, continued for a kilometre or so, then lay down under a bush. That night, it walked the perimeter of the forest and returned. And so it went for a day and a half, from four in the afternoon on 2 July until the morning of the 4th. What appears certain is that someone walked Cecil's collar around as if it was still attached to Cecil. Whoever did it was clever enough to mimic a lion's movements, buying crucial time perhaps for others to depart.

On 4 July, Brent Stapelkamp woke and, as he did every morning, checked Cecil's data to see where everybody's favourite lion was and what he was doing. 'I'm pouring coffee for my wife on the fourth of July,' he recalls, 'and I look—Cecil's collar sent me a point at 8 o'clock, then nothing at nine, nothing at ten. No data, no data, no data.' Although Stapelkamp didn't know it at the time, sometime after 8 a.m. on that morning, Bronkhorst's team disposed of, or destroyed, the collar. At first Stapelkamp just assumed that the collar had stopped working and thought to himself, 'I'll catch him when I see him next.' Stapelkamp had no reason to be suspicious—there were, after all, no lions on quota.

Three days later, on 7 July, nearly a week after Palmer had first shot Cecil, a safari guide turned up at Stapelkamp's house. Stapelkamp knew all the guides, who come from the same community as the trackers and skinners who work for the hunting companies. People had been talking, and the guide had heard that a lion had been hunted and killed. Stapelkamp knew straightaway that it was Cecil.

Stapelkamp and others pieced together Cecil's final movements and alerted the authorities. Within days, the story gathered momentum, spiralling beyond control, a conservation story custom-made for the media age. The bad guy—a dentist, no less—was a crossbow-wielding antihero who was easy to dislike. Our hero was a lion, the symbol of kings and already a star on social media. Told in breathless terms, and not always accurately, the story of their battle to the death waxed and waned, its outcome uncertain until the end. And our hero's death, when it came, brought a worldwide audience of millions to their feet. Jimmy Kimmel, Ellen DeGeneres and Ricky Gervais tweeted their support for Cecil to their combined following of tens of millions of people. On the one-year anniversary of Cecil's death, Leonardo DiCaprio tweeted to his 15.7 million followers using the #RememberCecil hashtag. Two years on, nearly 1.2 million people had signed a petition demanding justice for Cecil.

Away from the media storm and circus, the truth that emerged from the whole episode was far simpler. Cecil, the king of Hwange, was dead, aged twelve.

In the meantime there was a rather more personal footnote to Walter Palmer's shooting of Cecil. Jericho had seen hunters shoot Job, Judah and now Cecil, but that doesn't mean that Jericho understood what had happened, or was able to connect their disappearance with the terrible events that he witnessed. It is dangerous to ascribe human emotions to animals, but if you don't believe that animals are capable of emotion, consider this: for nights after his partner was shot and later killed, Jericho roamed the forest, roaring, calling for his friend.

Boundary Boys and Girls

Cecil's story—of the complex machinations of leonine life and the perils of trophy hunting—has echoes for many lions across Africa. But Cecil was only one lion and what was happening more broadly to the lions of Hwange—to Cecil's family and neighbouring prides—was just as important for the future of lions.

Imagine that a line runs parallel to the park boundary, roughly from the north-west to south-east, 10 kilometres inside the park. West of that line lies Ngweshla, the big prize for lions, and to its north the Kennedy *vleis*—long, low valleys, pride lands that belong to the kings of Ngweshla and that once belonged to Cecil. East of the line, between Ngweshla and the park boundary, lion home ranges are smaller, with names like Ngamo, Linkwasha, Mbiza, The Hide and Main Camp. To the west a single pride, to the east a patchwork of prides and coalitions. The latter are Hwange's B-team, a formidable lot and never far from danger.

Xanda

Xanda, born in May 2011, was one of the eighteen cubs that Cecil sired at Linkwasha between 2009 and 2012. He had quite the pedigree. Not only was Cecil his father, but his mother was Stumpy Tail, a grand old lioness of great character. Stumpy was the favourite lioness of many a safari guide and lion researcher. She lost her tail to a hyena in 2014, and throughout her life she held her prides together against enormous odds.

Xanda was eighteen months old when Bush and Bhubesi pushed Cecil from Linkwasha in late 2012, and Xanda was forced to flee, along with the rest of his cohort. An eighteen-month-old male lion in Hwange is usually twelve to eighteen months away from independence, from being able to make his own way in the world. When males arrive to take over a pride, and the former ruling male who has tolerated his sons flees, those same youngsters are in great danger: probably killed if they remain, barely able to survive on their own if they don't. We know from past experience in Hwange that the younger the male is when he leaves his natal pride, the less likely he is to survive.

If a young male manages to flee, there's a familiar pattern to what happens next. 'Usually the young lions in that area, if displaced by the incoming males, head along the southern and eastern boundary fence where they eat cattle and scavenge,' Brent Stapelkamp told me. 'After they serve something of an apprenticeship at the school of hard knocks, they come back to try find a territory.'

Xanda did just that. After fleeing Linkwasha, he shadowed the park boundary—sometimes tracking inside the park, sometimes outside it—until, perhaps following some indelible ancestral memory, he ended up where Cecil was first seen, down at Mangisihole. He survived, grew strong and then, for reasons known only to themselves, Xanda and his half-brother Sinangeni (born June 2012, also a son of Cecil), turned around and marched back on the land where they were born.

Things had changed while they were away. Hunters had shot Bush, leaving Bhubesi alone and vulnerable. Avenging their father, Xanda and Sinangeni cast out Bhubesi and became the rulers of Linkwasha. It would have been a stunning reversal, a real coup to be cheered for from the sidelines, were it not for one uncomfortable fact. The lionesses in Xanda's new pride, the lionesses with whom he would mate, were his grandmother, his mother and his sisters. And yes, in case you're wondering, Xanda knew exactly what he was doing: lions do indeed recognise their own mothers.

Xanda is far from the only male in Hwange's boundary prides to have mated with his kin. I talked with Jane Hunt about whether inbreeding could be a symptom of the wider social disruption that results from trophy hunting. Did the disarray that follows so much social upheaval cause Xanda to mate with his mum? Jane was rather polite in the face of my entirely unscientific question. 'I don't think we can simplify it to quite that extent,' she replied. But, she continued, 'I would say it's very definitely a contributing factor.'

If we can look past Xanda's lack of scruples, his time at Linkwasha was a success, at least in the beginning. He became something of a fan favourite among guides and guests of Wilderness Safaris. He was, after all, son of Cecil, a beacon of hope after the evil tragedy that had befallen his father. Like his father, Xanda was a good-looking lion. And he had cubs, lots of them. At the height of their powers, Xanda and Sinangeni held both Linkwasha and Ngamo, an impressive dual prize among the B-team.

But the tumultuous nature of life along the boundary would soon engulf Xanda and carry it all away. On 7 September 2015, Sinangeni wandered beyond the park boundary and into one of the hunting concessions, where he was shot by trophy hunters. He was just three and a half years old, well short of the accepted legal minimum for a trophy-hunted male. Xanda was present at the killing, but the hunter saw Xanda's collar and decided a young lion was preferable to risking another Cecil-like furore. Sinangeni's death would have catastrophic ripple effects elsewhere, but for now Xanda was left on his own. Soon enough, a pair of young males named Ngqwele and Butch seized Ngamo from Xanda. Without his brother as back-up it was one against two, and Xanda retreated north to Linkwasha to nurse his injured pride, his territory now cut in half.

And then history repeated itself. On 7 July 2017, exactly two years after the very week that Walter Palmer had killed Cecil, and barely 20 kilometres south along the railway line from where Cecil was killed, a hunter shot and killed Xanda, aged six. Perhaps he had crossed into the hunting concession to avoid the growing pressure from Ngqwele and Butch in the south. And maybe, if he had still had Sinangeni, he might have been more secure in his home range with no need to roam across the railway line. We will never know.

Without Xanda to protect them, the pride began to unravel. It didn't take long for Ngqwele and Butch to notice that Xanda was no longer around. In the weeks and months after Xanda's death, Ngqwele and Butch moved in, eager to mate with the pride females. Taking no chances, Stumpy Tail fled with Xanda's seven cubs. Three of these cubs were killed. While still on the run a year later, Stumpy was hit by a passenger train at 4 a.m. on 21 September 2018, aged eleven years and eight months. Thanks to Stumpy, four of Xanda's cubs—two females and two males—survived, barely. They now live in the shadowlands of other prides.

The Chiz Boys

Throughout early 2019, eighteen months after Xanda's death, the patchwork of territories that separates the park's frontier from Ngweshla was ruled by a succession of two-male coalitions. In the south, Ngqwele and Butch held Ngamo and Linkwasha. Immediately to their north was the narrow Mbiza region, a wedge of open country rich in palm trees and waterholes, which was held by Phobos and Damos. North again, Mopane and Seduli held The Hide. And then, as tends to happen along the boundary, everything changed.

On 8 May 2019, a train struck Butch, breaking his leg. The

authorities euthanised him two weeks later. Ngqwele was left on his own in charge of Ngamo and Linkwasha.

Six weeks later, on 23 June, Damos was trophy hunted, leaving Phobos in charge at Mbiza.

And on 31 July, hunters shot Seduli. Mopane was left without his partner at The Hide.

The boundary prides—their ruling males and, by extension, their cubs and the pride females who would defend them—were all suddenly vulnerable. For all lions in this area, the difference between survival and sudden death is a very fine railway line indeed. Drawing on long-term data sets, Dr Loveridge and his colleagues found in 2014 that prides along this boundary tend to be smaller and less successful, bringing far fewer cubs through to adulthood. For prides in the core of the park, 119 out of 130 cubs reached the age of one. In boundary prides, less than half of all cubs (53 out of 110 studied) survived to their first birthday. A cub born in the park's protected core had a 77.9 per cent chance of living to the age of two. On the boundary, just 19.2 per cent of cubs would make it that far. Ten out of 22 Hwange boundary prides studied had disappeared within just nine years.

As Jane Hunt told me, incoming males 'need a tenure of at least three years to bring their genetics through'. Or, to put it another way, a male needs to hold onto his territory and protect its females and cubs from rivals for long enough—ideally three years—for the cubs to grow strong and able to defend themselves against attack, hunt without the help of their parents, and disperse to find their own territories. Historically in Hwange, very few males along the boundary have lived long enough to meet this threshold, hence the low survival rate among cubs. That Cecil came close to three years as head of a pride is unusual, and it explains why six of his female offspring from his Linkwasha period survive. Most boundary cubs are not that lucky. At the end of 2019, Ngqwele was the only B-team male to have ruled for nearly three years, but he now had a second generation of young cubs to protect.

For all of the boundary males and their dependents, there were ominous signs. According to Jane Hunt, at least fifteen males, including a coalition of five brothers, were roaming the park's southern reaches. Should they sense an opening and march on Ngamo, Linkwasha and the rest, they could drive out the resident males and become the latest lions drawn from the park's interior to the boundary, sucked from safety and onto the front lines.

~

Change also came from the north. In July 2017, three male lions were caught on a camera trap in Chizarira National Park, more than 150 kilometres north-east of Hwange. It was an exciting discovery—the lions of Chizarira are in trouble and any evidence of lions there is considered to be good news. In March the following year, or perhaps it was April, the three boys from Chizarira headed for Hwange, passing through an area of farms and villages where livestock was the only prey. They didn't mess around. By May 2018, they were in Hwange.

Three males on the march was guaranteed to stir things up and, in July 2018, they evicted Mopane and Seduli from Main

Camp. Mopane and Seduli, who retreated to their smaller territory at The Hide, were no slouches. Main Camp is a popular tourist area, and Mopane and Seduli had developed a reputation for charging at vehicles and scaring the hell out of visitors; there was even a sign posted at Main Camp warning vehicles to keep their distance. Jane Hunt was rather fond of them: 'Mopane and Seduli were proper lions. They came from the middle of the park. They didn't grow up underneath a tourist vehicle when they were cubs. Many guides behave badly, and Mopane and Seduli were behaving as they should. If you've got bad enough manners to drive up too fast to them, or too close to them, well, yeah, they should be revving you.'

The Chizarira three—known by everyone as the Chiz Boys—were no soft touches themselves. As one experienced safari guide, Tendai Kedayi, told me, 'Being a nomad is tough. Some will go into the villages and cause problems and may end up getting killed. That's why when the nomads come in to contest for a territory, they're strong. If you survive as a nomad, it means you're a tough lion.'

Newly installed at Main Camp, the Chiz Boys began to wreak havoc. 'They gave us a hell of a run-around,' said Jane Hunt, 'because they caused a lot of conflict, killing cattle and stuff. They were seriously bad, because they went into *bomas* where herders keep their cattle.' An offshoot of the Hwange Lion Research Project, the Long Shields Lion Guardians—inspired by the Lion Guardians of Amboseli in Kenya and adapted to local culture and conditions—had been tasked with stopping just this kind of conflict. As well as educating local herders about better protecting their herds, the Hwange

Guardians chase off lions using loud *vuvuzelas*, with the aim of teaching lions to stay away from these human settlements.

For six months the Chiz Boys ran rampant. 'We chased these lions day and night,' remembers Liomba-Junior Mathe, the head of the team that oversees the Guardians. 'We had so many sleepless nights.' It got so bad that the head of National Parks in Harare ordered that the Chiz Boys be killed under Problem Animal Control (PAC) regulations.

Everybody out in the villages was on edge, and one day a call came in that a woman had been attacked in the forest. An attack by lions on a person could escalate things dramatically, so Mathe and his Lion Guardians made an urgent dash to the scene. There they found a group of shocked and confused women. Not everyone had seen the lion, but they weren't waiting around to do so. The woman who had seen it was bruised and bloodied and scratched all over—not from the lion, the Guardians were relieved to learn, but from running through the thick branches in terror and stumbling over the rocky ground. The Guardians took down the details and headed to the scene, to the exact place where the incident had happened. As always, they were subdued and nervous as they drew near on foot to a possible encounter with a lion. They looked everywhere and could find no trace of lion, not even footprints. Suddenly, something moved in the bushes. They moved in to get a better look. It was a calf!

With the Chiz Boys in the area, everybody was seeing lions, even where there weren't any to see. The Guardians reported back to the rather sheepish woman, who probably never lived down her mistake. Deep down, most locals would have understood. With lions about, it's better to be safe than sorry.

These lions were well practised in making their way unseen among the villages, and they stayed one step ahead of their pursuers: no one could find them to kill them as ordered. By the time they were tracked down, the Chiz Boys were down to two—the third male had been gored to death by a buffalo—and had settled down to rule over their pride near Main Camp. As long as the Chiz Boys stayed away from the villages, the authorities were willing to commute their death sentence.

The survival of these two important lions was an important victory for the Lion Guardians of Hwange. By chasing the lions and driving them away from human communities, never letting them rest or feed in communal lands, they saved lions and livestock. There were other success stories.

Mvuu, for example, was born into the Main Camp area in the pre–Chiz Boys era. Like any young male, he outgrew his pride and dispersed, looking for a land to call his own. For Mvuu, that meant shadowing the boundary with his brother, trying to avoid the pride males who would kill him on one side of the railway line, and trying to avoid the villages and hunters who would kill him on the other side. He dipped in and out of the park, tracking south, chased by Lion Guardians with their horrid *vuvuzelas* whenever he and his brother strayed too near to people and killed livestock. This continued for nearly two years.

Such is the lot of the dispersing young male, for whom it must be a terrifying, disorienting time. Mvuu's brother was snared and killed in one of the areas from which they were chased. I have seen the map of Mvuu's journey, the GPS plot points always close to the park boundary and any number of dangerous land-use areas outside the park. Mvuu made it, and now rules over a pride down at Josibanini. He is, said Jane Hunt, a splendid seven-year-old male.

But lion success stories in these parts invariably come with a warning. Xanda and Cecil showed us the dangers faced by lions who live just inside the boundary. The Chiz Boys and Mvuu remind us that a high price is paid by the people who live just outside it.

Ngqwele

While all of this was happening, Ngqwele, now without Butch, held on gamely to the Linkwasha and Ngamo prides, far enough away from the Chiz Boys to be safe for now. It helped that his neighbour to the north, Phobos, had also been recently left alone and was fully occupied just holding his ground at Mbiza.

Ngqwele is not Hwange's loveliest lion. His mane has the lion equivalent of a receding hairline and, in this, he is the spitting image of his father, Bhubesi, who had driven Cecil from Linkwasha way back in late 2012 and early 2013. Good looks may not be on Ngqwele's side, but he has earned the respect of many guides as a lion to be reckoned with. He is known to take down buffaloes on his own, and he works hard to look after his considerable brood. At Linkwasha he is protector of two adult females and six cubs, while down at Ngamo are four adult females with three cubs and subadults, as well as a new batch of six cubs. One day when I was in Hwange, he was seen by guides near Linkwasha. A day later he was down at Ngamo. The following night he marched 10 kilometres back to

Linkwasha where he woke up guests—including me—at the tented camp.

That all of this should take place at Ngamo makes his success even more impressive. In the rainy seasons of December and January, and April and May, Ngamo is perfect lion habitat and much sought after, an arid land turned green that draws vast herds of wildebeest and buffaloes from elsewhere in the park. Prior to that, in most Septembers and Octobers, elephant carcasses litter the pan and surrounding woodland at Ngamo, providing ample food for those who hold the territory.

But Ngamo is a fragile paradise. On one hot and windy September afternoon, I stood at the centre of Ngamo and marvelled at its bounty for lions. As I did so, a freight train appeared and rumbled past.

'If you look at Ngamo,' Jane Hunt said, 'you've got hunting and the railway line and possibly snares on one side. Down another side you've got hunting, snares, PAC, cattle, problems with people. It's pretty hectic.'

Dr Paul Funston of Panthera agrees: 'It's a perfect storm. You've got the perfect habitat in the park. Just across the railway line you've got hunting to the one side, and to the south or east, you've got community lands. Ngamo is a tragic corner.' If he goes west, Ngqwele should be fine. If he goes east, south or north, he's in trouble.

In all of this, Ngqwele is very much a lion of his time and place. Buffeted by the social disruptions that afflict boundary prides, he has adapted and survived, even generating enough stability amid the chaos to bring cubs to near adulthood. From one perspective, Ngqwele offers hope; he is a testament to the adaptability of the lion as a species. The stark lesson is that lions must adapt or die. Just being a normal lion doing normal lion things is no longer enough.

But Ngqwele is just as much a symbol of the rampant dislocation that conflict along Hwange's border has wrought. First, like Xanda, Ngqwele has mated with his mother, his sister, his aunts and probably his own grandmother. He was born to Bhubesi in the Linkwasha region in April 2014, and, after dispersing, he returned to the pride of his birth.

Second, like any male lion, Ngqwele no doubt killed the cubs of other males when he moved in and seized control of Ngamo and Linkwasha. But what he has also done in recent times is kill his *own* offspring. It is not unusual for a male lion to drive his own nearly adult sons away from the pride—it happened to Ngqwele as a subadult. But he has already killed two of his own sons, neither of whom could possibly have been old enough to leave the pride and become independent; one of his sons, Zandi, was just two and a quarter years old. Inbreeding and murdering members of your own family are hardly glowing advertisements for the health and stability of Hwange's boundary lions.

There are other behaviours happening down in Ngqwele's lands that simply don't happen in more settled lion lands elsewhere. As soon as numbers grow in the boundary prides, they tend to split up, reducing their size as targets and minimising competition for food. This is not unusual in lion society where lions live in close proximity to human communities—the same thing happens in Amboseli. What hasn't been recorded in other places is 'appeasement mating', the process by which

Hwange boundary females will send one of their number to mate with an incoming male and distract him while the others hightail it with the cubs. Or entire cohorts of cubs are dumped with a single female while the remaining females head off to mate with the new male, sometimes never to return. Or, in what Jane Hunt somewhat unscientifically describes as 'double declutching', females begin mating with pride males while still caring for juvenile or subadult offspring.

These unusual aspects of lion behaviour may be ingenious survival techniques, an example of what Dr Loveridge and his colleagues describe as lions showing 'a high degree of flexibility and opportunism, particularly in human-disturbed environments'. But they could also be desperate acts, with long-term consequences for lion society and survival that we don't fully understand. One thing that is certain is that most of it would not be happening were there not so much conflict on their doorstep.

Hwange is not some marginal lion land where things like this don't matter. Hwange contributes between 500 and 700 of the nearly 2000 to 3500 lions thought to be living in the broader Kavango Zambezi Transfrontier Conservation Area (KAZA). The KAZA lions are one of the most important lion populations in Africa. According to Panthera's Paul Funston, KAZA is 'the largest conservation landscape in Africa' and 'one of the most important regions for lion conservation in Africa, probably the most important in southern Africa . . . There are less than five such large functional populations left on the continent, probably only three.' In such a significant population, we brush aside these symptoms of social disruption at our peril.

All the action around Ngqwele can be exciting to watch. But lions thrive on stability, and it's difficult to avoid the fear that Ngqwele himself may not be around for much longer. In 2019, when I asked one expert familiar with Hwange's lions whether Ngqwele had strayed into the hunting concessions, his answer was hardly reassuring. 'He's in there all the time. It's part of their range, all of that forest. The hunters wanted to shoot him this year, but they decided not to. They've already shot Seduli and Damos this year.' Being on quota, the hunters were not able to shoot any more lions; Ngqwele, five years old at the time, was also underage for a trophy-hunted lion. 'The hunters that were in there with lion on quota to hunt this year chose not to, because of Ngqwele's age and because he's with a pride.'

Ngqwele was safe. But all bets would be off when he turned six—the minimum age used by many hunters—just in time for the hunting season in 2020. And then a new round of chaos could begin.

The Ngamo Urchins

If the male lions create much of the excitement, seizing and defending territories then too often dying in a blaze of glory, it's the females who pick up the pieces long after the males have gone. The stories of Cecil and Xanda, of the Chiz Boys and Ngqwele, evoke the sound and the fury in the lives of Hwange's lions. The stories of the Hwange females make us wonder if there is any future at all.

It would be easy to dismiss the stories of Linkwasha, Ngamo and all the boundary lands as atypical within the broader Hwange lion population, and within other lion populations in Africa. But they're not. As studies by the Hwange Lion Research Project over many years have shown, what happens along the boundary carries the long reach of human intervention—trophy hunting, PAC, human–lion conflict, snares, even the railway line—into all but the deepest core area of the park. Almost every time a boundary pride loses its resident male or males, the vacuum effect draws another male or two into the danger zone and pride life begins a new round of dislocation. The Hwange Lion Research Project's own data shows that only around one-third of the park is safe from what happens on the boundary. And what happens there can be catastrophic.

~

On 6 December 2007, Jane Hunt came across a young female lion down at Ngamo. The two-year-old lioness was in poor shape. She had been badly beaten and was deeply scarred around her rear; her ears were swarming with ticks, and one sagged flat. Just a month before, vets had darted her so that they could remove a snare from around her neck. She was wandering Ngamo with three of her siblings who were similarly ragged. Hunt remembered her well: 'I collared her as an urchin. She was a street child. She was tiny. Several of the siblings were caught in snares and died. The mom was caught in a gin trap and attacked a guy in Tsholotsho.' The mother didn't survive.[3]

A group of two-year-old subadult lions was cast adrift in one of the busiest lion areas of the park with no one to protect them—their days should have been numbered. So many lions in Hwange find themselves in similar situations, and happy endings are few: they're either killed by other lions, or they stray into communal areas and end up getting killed by people, or they disappear into the wild, fate unknown. Ngamo, then as now, was crowded with lions, including resident females and territorial males, and the urchins' survival depended on the complex tasks of ducking and diving, of staying away from other lions yet killing prey often enough to survive. Such complex manoeuvring is beyond many young lions, and Jane Hunt wasn't hopeful for their prospects. 'They shouldn't have survived,' she told me. 'They were too young to be on their own. Theoretically they were in the pantry, and theoretically they could scavenge. But very early on, I caught them, this girl, having killed a full-grown eland!'[4] The surprise in Jane's voice was still strong twelve years later. 'So these youngsters were quite impressive. They were capable, and against all odds they managed.' The pride became known as the Ngamo Urchins.

Led by this young female—NGAaF1, as the original Ngamo Urchin was called in the data records of the Hwange Lion Research Project—the pride stayed at Ngamo. It would be wrong to say that they flourished, as we shall see. But their leader managed to survive, and she and her pride mates rarely killed livestock—a remarkable feat considering her pride's territory bordered an area hectic with farms and cattle. 'She did incredibly well,' Jane Hunt remembered. 'And they were super well-behaved. They had obviously learned from their early days. She was a very good girl.' She did cross the park boundary from time to time, probably to escape aggressive males. There she lay out in the open and did nothing to annoy the villagers.

A succession of males—including Xanda, Bush and Bhubesi—all passed through. She mated with all of them and had cubs with most of them. These males would each hold Ngamo for a time, but never long enough—never for the three years needed to allow a cohort of cubs to survive. The Ngamo Urchin had five, maybe six litters of cubs. Some were killed as a result of human–lion conflict. One was hit by a train. Others, cubs who never lived long enough to make it into the records, were victims of infanticide from incoming males. Only two daughters, born in December 2009, survived to adulthood. The pride's membership was in a constant state of flux, but the core around which the chaos swirled was four lionesses: the original Ngamo Urchin, one of her sisters and the two daughters born in 2009.

Males may garner the headlines as protectors of prides but their protection is a snapshot in time. Prides endure over time and generations because the females are the glue that holds everything together. It's a point that Jane Hunt was keen to emphasise. 'The old girls are the ones who rule,' she told me. 'It's like any family, whether lion or elephant. Those are the aunties that discipline, that hold it all together. They also know where to go in hard times. This is what we're talking about with the stability of the pride, the knowledge of the pride, the body language of the pride, and the knowledge of knowing how to react, how to discipline, and how to keep everything in line, because if you end up with a bunch of youngsters you get *Lord of the Flies*.'

With everything that was happening, these Ngamo Urchins were extraordinary lions. For nearly nine years they survived in one of the most dangerous areas of the park. They stared down the recurring tragedy of losing cubs. And they chose to mate with each new conquering male in the hope that he could help carry their cubs into adulthood and ensure that their bloodlines outlived them. Through it all, the original Ngamo Urchin—the pride's matriarch—held everything together. Somehow, they survived without leaving the park and killing livestock which, compared with the danger they faced from each new male, would have been the easy way out. Cecil may have captured the world's imagination. Ngamo Urchin was the real hero.

~

Before we follow the Urchins in their desperate battle for survival, it's time we addressed what some lion researchers consider to be the elephant in the room: hunting, and its role in causing conflict between lions and people.

According to the published data of the Hwange Lion Research Project, trophy hunting has been responsible for 10 per cent of deaths among the studied lion population. A further 40 per cent is caused by human–lion conflict and/or what researchers call anthropogenic factors—snares, PACs, revenge killings, perhaps even trains. The remainder—half of the lions in the study—die natural deaths (killed by other lions, mortally wounded while hunting, killed by disease and other non–human-related causes). All of this falls within expected levels and is considered sustainable provided that only mature adult males are shot, quotas are based on detailed population estimates, and so on.

But at some point, it became apparent that these figures didn't tell the whole story. After talking with those on the front line of the conflict—the villagers who live along the park boundary—the lion researchers discovered that there was a discernible spike in lions killing livestock, and lions being killed in retaliation, in the months after a pride male was trophy hunted along the boundary. 'So what was causing them to leave the protection of the park and die at the hands of people?' Dr Loveridge asked. 'When I looked at the data more closely, an interesting pattern emerged. Spates of livestock killing, and consequently the deaths of adult females, frequently occurred when the territorial male had been killed, often by trophy hunters.' Or, as Brent Stapelkamp put it, hunting 'forces them into areas where they find more snares, where there's less prey, where there's more conflict with people . . . There's two demographics that usually kill livestock—subadult males and lionesses with cubs. Both of these demographics are pushed out of the protected area when there's a change in the tenure of the territorial males.' Taking these suspicions as their cue, the team collared a number of lions and began to test the hypothesis.

What they found was staggering and could prove controversial: perhaps hunting just outside national parks is not sustainable at all.

The study results found that most of the 40 per cent of lion deaths attributed to human–lion conflict could actually be traced back to hunting. Put another way, in Hwange at least, up to half of all lion deaths were driven by trophy hunting. The hunter who pulls the trigger kills one lion—this is the 10 per cent directly attributable to trophy hunting. The chaos that results from this killing creates all manner of social disruption—infanticide, the splitting of prides—that sends lions scurrying into confrontations with human beings as lions flee the park. This in turn drives human–lion conflict, causing more lions to die—the further 40 per cent who die at the hands of humans. If trophy hunting causes, directly and indirectly, half of all lion deaths in the study area, then trophy hunting is no longer sustainable, if indeed it ever was.

Across Africa wherever trophy hunting occurs, these findings, when published, will be political dynamite.

~

The Ngamo Urchin could only keep the chaos at bay for so long, and in 2015 her world finally began to unravel. In January that year, Bush and Bhubesi killed her sister, the one who had been with her from the beginning. In mid-2015, Xanda and Sinangeni ruled at Linkwasha and Ngamo. Xanda had mated with the Ngamo Urchin and the other females—what genes their young cubs must have had, with their direct bloodline from Cecil, Xanda's father, and the original Urchin, one of the park's great matriarchs! On 7 September 2015, Sinangeni was trophy hunted, as we have already seen. Xanda, suddenly alone, began to lose his grip on Ngamo, which didn't go unnoticed by other males waiting in the hinterland for their chance to move in.

The Ngamo Urchin had seen it all before. But this time something snapped inside her. Having never strayed into the villages and farms, she finally decided to take her chances among people rather than risk yet another cohort of cubs dying in the jaws of incoming males.

At first, what she did was textbook lioness behaviour. It has long been an article of faith among Hwange lion mothers that they and their cubs have less to fear from people than they do from marauding male lions. 'We all know that when a pride male is challenged by other males and loses,' Brent Stapelkamp told me, 'he gets killed or chased away, and there's infanticide—the cubs get killed. On paper, it sounds clinical and that's what happens. But the reality is that lionesses are typical mums. So if they lose their sire to a trophy hunt or to a legitimate fight, they don't just hang around and wait for the cubs to die. They either fight and try to defend, in which case they most likely get killed, or—more often than not—they take the cubs and run.'

But their choices were limited, as Stapelkamp pointed out. 'Where, in the modern African context where we understand that even our biggest national parks are not big enough, where do they go that there are no other lions that are going to kill their cubs? They leave the park. And where do they find themselves? Smack-bang amongst people, where there's nothing to eat but livestock.'

And so it was that the Ngamo Urchin gathered up her pride, or what was left of it—herself, her two adult daughters, and seven cubs (including two eight-month-old cubs to Xanda) and subadults—and left the park. On 2 March 2016, they killed four donkeys. Four days later, the Urchin's two adult daughters, at seven years old, were PAC'd—shot as problem animals. Later the same month, male lions killed one of her cubs to Xanda, close to the park boundary.

Then she did something that no Hwange lion had ever done before. Knowing that Ngamo was unsafe because of the males, and apparently understanding the risk of staying close to the villages, she took what remained of the pride and ran 23 kilometres, in a straight line, to a forest area called Mapengula where there were very few villages. From around 8 kilometres beyond the park there would have been no more wild prey, so they continued to prey on livestock. But being so far from the park and away from more densely populated areas bought them some time. It was a stroke of genius. Every few weeks, Ngamo Urchin would make the long trek back to Ngamo to check whether it was safe for them to return. She would stay for a day or two, then return to Mapengula.

Sadly, Ngamo Urchin's heroic defence of her pride only delayed the inevitable. Three of the cubs and subadults were snared and killed out near Mapengula. In October 2016, the last cubs were snared in retaliation when they returned to feed on a donkey they had killed near the village of Gwenga as they made their way back to the park. One young male, a year old, was saved from the snare, treated by park officials and then, inexplicably, released at Nyamandlovu, close to Main Camp; he was quickly killed by the resident males.

No one knows exactly what happened to Ngamo Urchin: her collar was never recovered. Most likely this tragic lion queen was snared and then killed. For all of her heroics and years of good behaviour, and despite her extraordinary tale of survival, not a single member of her pride, not a single carrier of her bloodline, survived. It had all been for nothing.

Cecil: What Happened Next

The shooting of Cecil by Walter Palmer set off a firestorm. For those on the ground, it was an extraordinarily difficult time.

Hunters in Hwange, including those who denounced the illegal hunt, received death threats. So did Andrew Loveridge and Brent Stapelkamp; 'I had to sleep with my .375 loaded next to my bed,' remembered Stapelkamp. 'We were fully expecting any night for a big Land Cruiser to come into our yard. It was a terrible time for my family and I.' The controversy over Cecil also stoked tensions between the Hwange Lion Research Project and the national park authorities. Both opened bank accounts to handle donations after the Cecil story went viral. More than a million US dollars flooded into the coffers of the lion project's Oxford headquarters, and most of this was spent transparently on lion conservation projects inside Zimbabwe. In contrast, no one donated to the Zimbabwe's National Parks authority.

The fallout from Cecil's death went beyond money. 'Cecil brought an awareness of the species, and of conservation at its best and at its worst,' said Jane Hunt. 'Obviously it brings attention to the plight of lions. For me, the most amazing thing that happened is that it showed it's not just conservationists fighting this battle.'

Or, as leading lion expert Craig Packer told me, 'In some regards it was a sideshow, because it was just one animal.

It happened to have an appealing name for Western audiences. There are plenty of other lions who have been shot before, during and after all of that that we didn't hear about. I guess there might have some perverse sort of pleasure about pointing the finger at a dentist. It's certainly a sideshow in terms of what goes on.' Yet it did, he added, push some organisations into re-evaluating the practice of trophy hunting. 'That firestorm of interest certainly had the consequence of making regulatory agencies more sensitive to the whole question of whether sport hunting should continue as it has done or whether it should be reformed. That could be seen as a very positive outcome.'

Around Hwange, change had begun even before Palmer killed Cecil. The government was already working closely with conservation groups like Panthera and the Hwange Lion Research Project to keep hunting at a sustainable level. In decades past, 36 Hwange lions were available to hunters on quota every year. By 2019, the number was closer to six.

Yet at the end of it all, Cecil's death merely made public what many conservationists already knew: lions are in trouble. As Brent Stapelkamp put it, 'This is lions. We are losing them to the point where they will only exist in the strongest of protected areas in a decade's time.' In this, and in many things, nothing has changed.

Cecil was just one lion and what matters is the whole population. Together, the tales of Cecil and other Hwange lions paint a bigger picture of the perils confronting lions across Africa. In this, Hwange is a touchstone for the continent and its lions, and we ignore its lessons at our peril.

For all that, it is the story of Cecil that captured the world's imagination at a critical moment, and he remains the tragic, slain, undisputed king of Hwange. Safari guides, lion researchers and just about everyone else in and around Hwange still field questions about Cecil—his life, how he was killed, and where it happened. People talk about Cecil as if they knew him, or as one might talk about a celebrity who died before his time. They want to see where he lived, as if that will make them feel a part of the biggest wildlife conservation story of our time. Close to Linkwasha, guides routinely stop at a bleached buffalo skull. 'Cecil killed this buffalo,' they tell everyone, and we all take a picture and nod knowingly.

And, like all great stories, we all want to know what happened next.

~

After Cecil's death, Jericho returned to Ngweshla alone. There were concerns that he might harm Cecil's seven cubs, who were mostly between one and two years old at the time. After all, the coalition that was Cecil and Jericho was one of convenience, not based on the rule of shared blood that ties most male coalitions together. But, as Jane Hunt said, Jericho was always a lover, not a fighter, and he just didn't have it in him to kill the cubs. His heart was no longer in Ngweshla, and he spent most of his time up on the Kennedy *vleis* with another group of lionesses. He no longer mixed with Cecil's females—the Spice Girls—or Cecil's cubs. But he tolerated them and did enough to ensure that no other males moved in to take the territory.

Life continued in this way for more than a year until, on 25 October 2016, fifteen months after Cecil's death, Jericho curled up under a tree and died. He was nearly twelve and a half years old, and he was the only male member of his family to die a natural death, to die of old age. The placid lion, the one with no stomach for a fight but plenty of stomach for food, outlived them all—his ferocious father, his darker-maned brothers, his famous sidekick to whom he deferred. And he had seen it all—his father's death in battle, the slaying of his two brothers, the slaying of Cecil, the vicissitudes of winning and then holding a prized lion territory, even the snare around his neck and the years in the wilderness. Between them, Cecil and Jericho sired as many as three dozen cubs. As Brent Stapelkamp saw it, 'Jericho's life was an expression of all that a lion sees in modern Africa.'

With Jericho gone, that old opportunist Bhubesi moved into Ngweshla, raising the stakes even higher for Cecil's family; most of Cecil's cubs were still too young to disperse. Jericho had at least been Cecil's coalition partner. Bhubesi had no such connection. Stories differ as to what happened next. Calvet, the safari guide from Somalisa, remains convinced that the Spice Girls, clever to the end, mated with Bhubesi by pretending to be in oestrus, and continued mating with him until he could mate no more. 'They kept him busy for three days until Bhubesi couldn't even make the mating sound. Bhubesi was so quiet, so tired and hungry. And probably the girls said, "If you accept our kids, we will allow you to stay with us."' Jane Hunt suspects that it was the rather unimpressive manes of Cecil's sons—'They looked like girls'—that led

Bhubesi to feel no threat from these upstarts. 'For whatever reason,' she remembered, 'he didn't kill Cecil's sons. It's just bizarre. And he didn't kill the females. Then he mated with the females, and he was just living among all this rabble. He wasn't great to them, but he tolerated them. They all lived. He didn't kill any.'

Bhubesi stuck around long enough to mate properly with the Spice Girls who, inevitably, gave birth to Bhubesi's cubs in early 2018. As a lone lion in charge of the prized Ngweshla pans, Bhubesi was always a stopgap pride male, and it would be his last posting in a long, peripatetic life that was never far from the major upheavals that swept through Hwange lion society. In the middle of 2018, two strapping male lions, Humba (born May 2013) and Netsayi (born December 2014), half-brothers, marched into Ngweshla. Bhubesi had been around long enough to know that his time was up, and he surrendered his territory for the last time and without a fight. He was last seen in late 2018, out in the backblocks beyond the Kennedy *vleis*, in poor condition. His body was never found, and he was most likely eaten by hyenas.

When I asked Jane Hunt who she considered to be the next Cecil, she didn't have to think for long: Humba and Netsayi. They had come from deep in the park, from good wild lion stock—Mopane and Seduli were their fathers—and, now that they held Ngweshla, they had no rivals.

Their arrival had, of course, sent the Spice Girls scurrying, less for their connection to Cecil—whose cubs were now old enough to survive on their own—but because their cubs from Bhubesi were barely six months old. They fled north, following the route taken by Cecil before his death—up through the Kennedy *vleis*, past The Hide Safari Camp, and then out of the park and into the Gwaai Conservancy forest—before looping back into an area known as Madison, close to the Linkwasha airstrip. There they came and went, the two old girls (Cecil's former 'wives'), four Bhubesi cubs (one female and three males) and—perhaps he just couldn't bear to leave his mum—one of Cecil's four-year-old cubs. The group achieved a certain notoriety when they raided a small plane that had been left overnight on the airstrip with the door insecurely fastened; they climbed aboard, dragged out some suitcases and cast some unfortunate traveller's underwear all across the airstrip. Ever since, they have been known as the Baggage Handlers.

Left behind at Ngweshla with Humba and Netsayi were four of Cecil's daughters, now of age. They duly mated with the new males and had six cubs, Cecil's grandchildren.

And that's how I found them in September 2019.

~

Close to the end of my time at Linkwasha I headed out to Madison, hoping to see Cecil's son, or two of the Spice Girls, or the whole group of seven that had been camped out at Madison for weeks. Madison is perfect lion habitat—a long, open valley fringed with forests of teak and acacia, and a large waterhole at the valley's low point. Two buffalo skeletons suggested that the Madison Seven were doing well enough for themselves. My guide, Tendai Kedayi, had, on another occasion, watched the seven outlast a mother elephant who was determined not to abandon a calf that had died of exhaustion; they gathered around to feast when she finally fled.

Emerging from the southern flank of the forest, SPIGF3, the only one of the Spice Girls to which a name never attached, walked out into the open and began to roar. A beautiful lioness with barely a mark upon her, she was alone and magnificent, a strong, vital connection to the days when Cecil had ruled this land. A herd of roan antelope—safe, hundreds of metres away—approached the waterhole warily, while a young male elephant came closer to look at SPIGF3 and at us, then tossed his head and hurried away. So regularly had the Madison Seven been seen here together that it was unusual that she was alone, that no one answered her call. We wondered if she was not calling to her sister and the cubs but to other lions, to males who would rule this land.

It became apparent the next morning why the others had remained hidden. That morning, out upon the valley floor, sat Humba and Netsayi—regal, lords of their world. No other creatures dared to walk upon the plain with these two present; Humba sat above a warthog burrow, peering down it lazily from time to time, terrorising the inhabitants that were no doubt trapped inside. Humba and Netsayi were perfect lion specimens, with luxuriant manes and barely a scratch on them—they had claimed Ngweshla without a fight. They didn't flinch, didn't even open their eyes whenever we turned on the engine or spoke in our open vehicle just metres away.

Even so, there was no doubting their ferocity. They had already proven their pedigree as Hwange lions capable of bringing down elephants. And their presence at Madison defined the place that crystalline morning in a way that only a lion can. If they found Cecil's four-year-old son here, or the other three male cubs of Bhubesi, they would kill them without hesitation. Perhaps SPIGF3 had known this the previous night—that the cubs had gone and that it was time for her to return to Ngweshla and rejoin her daughters in what was once Cecil's pride. Not long after my visit, she did indeed return to Ngweshla, probably to mate with Humba and Netsayi.

But Humba and Netsayi also had a softer side that will no doubt endear them to the safari visitors who come seeking the new Cecil. Netsayi is, according to Jane Hunt, 'a real sweetie'. 'Netsayi is called Netsayi because when I first put the collar on Humba, Netsayi wouldn't leave his mate alone when he was trying to wake up,' Hunt recalled. 'He was chewing his ear, chewing his collar, chewing everything. And that's what Netsayi means—'pain'. He was like, "What's wrong with you? What are you doing?"' You can imagine how bereft Netsayi would be if he ever lost his partner.

Could that happen? 'You never can tell,' said Hunt. 'You would think there would be enough to keep them at Ngweshla.' They should be safe. But Cecil, and Job and Judah before him, were all killed while ruling Ngweshla; where we saw Humba and Netsayi at Madison is just 12 kilometres in a straight line from the railway line and the hunting concessions. Like all of the Ngweshla males before them, they are within a danger zone from which they could be sucked out and shot.

As Cecil taught us, no lion is safe.

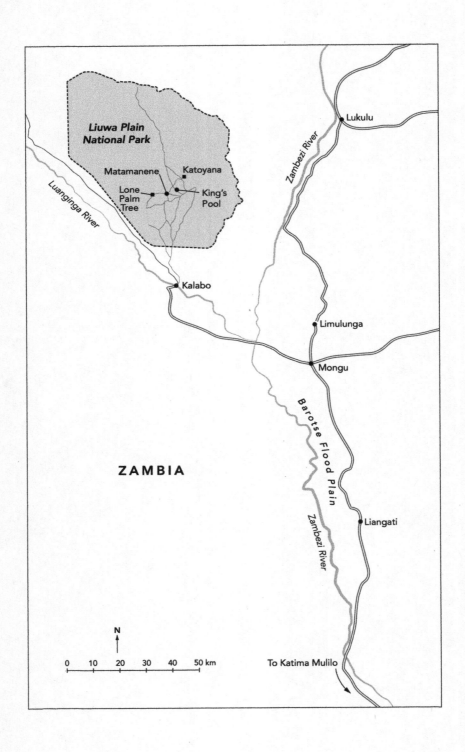

3

The Last Lion of Liuwa

Zambia, 2016

No one knew more about the lions of Liuwa than Jakob Tembo. In 2004, twelve years before I knew him, Tembo was already a veteran of Liuwa's troubled story, and the park he was responsible for protecting, Liuwa Plain, was a national park in name only. Since the 1970s, guns had flooded across the border from the civil war in neighbouring Angola, and a wildlife apocalypse wrought its fury upon this remote corner of western Zambia. When the Zambian government, its coffers empty, walked away from Liuwa sometime in the 1990s, the poachers moved in and rangers like Tembo watched from the sidelines, helpless, as men with guns wiped out all the animals.

At the turn of the 21st century, the park's boundaries meant as little to the armed men who wandered with impunity as they did to wildlife bureaucrats in far-off Lusaka, Zambia's capital. If Liuwa registered at all in government circles, it did so only long enough for local officials to issue hunting licences for those who would empty Liuwa of its animals.

Tembo remembered when Liuwa had been a true lion land, home to six prides who divided the park between them—a privileged pride around King's Pool, another from Matamanene in the heart of the park, the bad-ass pride from the north. A pride of twenty lions, ruled by a formidable coalition of five males, had once reigned supreme. On one occasion, before Liuwa went to ruin, Tembo had watched sixteen lions feeding on a single carcass.

But that was before, and when, in 2002, someone finally decided that Liuwa was worth saving after all, Tembo was one of the first to go back in. What he found confirmed what everyone had feared: there were hardly any animals left. An occasional, frightened wildebeest. A wild-eyed zebra. Wary hyenas. The question remained: was it all too late?

Liuwa stood on the brink of an all-too-common African story where wilderness areas disappear and wildlife becomes extinct. Once gone, they are gone forever.

And then, in 2004, something improbable happened: Tembo and his companions heard a lion roaring in the night. In desperate need of good news, they set off in pursuit, although they expected that any surviving lions would not want to be found.

Close to midnight, some weeks into his search, Tembo came across a lone lioness crouching in a sandy ditch. She was near King's Pool, close to where local Barotse kings had for centuries come to hunt. It was a wonderful moment, confirmation that all was not lost—a symbol that Liuwa had turned a corner at last. Tembo was with South African filmmaker Herbert Brauer, who was documenting Liuwa's halting re-emergence, and over the days that followed they tracked the lioness, hoping that she would lead them to others. They watched her hunt wildebeest, watched her do battle with spotted hyenas, watched her on her daily quest for food. And they listened in vain as her calls went unanswered.

Although greatly outnumbered by Liuwa's warring hyena clans, this lioness was a formidable presence on the plains. On one occasion, Tembo and Brauer watched as she killed a hyena clan matriarch, consigning her enemies to weeks of chaos as contenders to the hyena throne battled for supremacy. Whichever matriarch took over, she would know not to mess with this lion. 'One hyena was not equivalent to her,' Tembo told me later. 'Six hyenas were not equivalent to her.'

Tembo and Brauer followed her every day. Lions elsewhere quickly become accustomed to the presence of vehicles. Whether in South Africa's Kruger National Park or Kenya's Masai Mara, lions barely stir when a vehicle draws near, so familiar are they with the intrusion. In Liuwa, where trophy hunters and poachers had pursued lions with such ferocity, Tembo expected the lioness to take longer than usual to make her peace with their presence. Yet barely a week after Tembo and Brauer began following the lioness, they noticed something strange: when they weren't following her, she followed them.

There was nothing sinister about her movements. She wasn't stalking them. She wasn't seeking cover to mask her approach as she would if she was hunting. She walked in plain view.

Also—and this was when things started to get really strange—her behaviour didn't change when they left the vehicle.

It is generally understood among conservationists and safari guides that lions, once accustomed to cars, see vehicles as just another animate object in their world. People inside these vehicles, so the argument goes, are not in any danger as long as they remain within the vehicle and avoid any sudden movements that change the vehicle's shape in the eyes of a lion. As any half-decent safari safety briefing will tell you, if a person leaves a vehicle, all bets are off.

On one occasion in Botswana, I happened upon a male lion sleeping under a tree, perhaps 25 metres away. He watched my 4WD approach but he barely moved. Curious to test this theory about lions reacting differently to people once they leave their vehicle—just to see his reaction, just to understand mine—I opened the car door and placed a single boot into the dry grass that crackled underfoot. Instantly alert, the male lion snapped to attention, his eyes widened, and his muscles tightened as his body readied itself to spring. I clambered back into the vehicle with indecent haste.

In this context, the behaviour of Liuwa's lioness was odd.

And then, a few weeks after first encountering the lioness, Tembo awoke in his tent to the uproar of hyenas laughing. It is a sound both thrilling and profoundly unsettling, this manic laughter without mirth, as if the hounds of hell have streamed from the gates to cast terror into the night. It is the sort of sound that makes you pull the sleeping bag more tightly around you. Tembo, of course, took the pragmatic view that it was a filming opportunity that Brauer would not want to miss.

He rose quickly and prepared for the short walk to Brauer's tent.

I have shared a campsite with Tembo and he is not one to fear the night. He walks through the darkness not so much fearlessly as without concern, taking neither risks nor precautions as to his safety. He understands that he has little to fear from wild creatures. He also knows that that terrible accidents can happen—at fifteen he survived the bite of a spitting cobra. On one occasion, while hyenas circled us as we ate, he barely stirred. I was unnerved, but Tembo carried on with the calm fatalism that I have seen in so many Africans who have little choice but to move around the bush at night.

And so it was, too, on that night when the hyenas' laughter haunted the plains. Tembo unzipped his tent, took three steps and froze. There before him, he later told me, was 'something darker than the darkness'. He knew instantly it was the lioness.

In a world where lions and people do their best to avoid getting too close to each other, Tembo had never before been this close to a lion while on foot, let alone in the kind of darkness in which lions routinely cloak themselves to hunt. He was entirely at her mercy. Before Tembo could decide on his next move, the lioness rose and moved quietly out of his way. His heart racing, he passed where she waited. She turned to follow. In Tembo's telling of it, the lioness, unlike Tembo himself, was 'very relaxed'.

Not long after this, Tembo was filling a container at a waterhole when he turned to see the lioness standing on the rise behind him, just 2 metres from where he stooped. She could have killed him in an instant. Once again, she moved to one side to let him pass. 'Then she started following me. Herself behind, myself in front,' Tembo remembered. He kept walking.

Soon after, as Tembo cooked by the fire, the lioness returned to watch him. As she came closer, she rolled over as if playing, rubbing against the nearest tree, then rolling again, legs in the air. 'Like a cat,' said Tembo, shaking his head as he remembered that strange night when an otherwise wild lion chose him as her companion. 'Like a domestic cat.'

Later that night, Tembo woke in his tent to feel the earth moving. Momentarily disoriented, he wondered if it was an earthquake. And then he understood: the lioness was lying up against him, with only tent canvas between them. He swears to this day that she was purring.

Over the months that followed, the wild lioness became a regular visitor to the camp and an uncomfortable truth dawned on Tembo. There were no more lions to be found. This lioness was Liuwa's very last lion. And she was lonely.

～

The story of Lady Liuwa is a strange tale. Here was a lioness revered by the local people, a lonely lion who chose a human being to be her companion. It was, I would learn, a darkly spiritual fable of tribal traditions under threat and of a people's ancient ties to the natural world. And it was an example of that old pan-African legend: that men and women can turn into lions.

Humans have long believed that we share a special bond or

understanding with lions. Pliny the Elder, a Roman philosopher of the first century AD, assured his readers that lions were the only wild creatures to show compassion. In *Aesop's Fables*, Androcles takes refuge in the lion's den and they become allies for life. In the Muslim tradition, the lion was friend to the Prophet Mohammed, protector of Ali and Hussein. Throughout Western history, lions have embodied royalty, strength and noble values. But it is in Africa that people tell of human beings turning into lions, stories told by peoples from the San of the Kalahari Desert to secret societies in the continent's north-west. Few such tales have ever been told beyond the communities themselves.

The opportunity to tell Lady Liuwa's story was reason enough to come, to be sure. But it helped that the backdrop to this story was hidden away from the world in Zambia's extreme west. Gloriously remote, a part of Africa little known to the outside world and not really on the road to anywhere, Liuwa was my sort of place.

East of Liuwa, the national parks and safari circuits of Zambia are so far distant that they feel like an alien land. Travel west from Liuwa and you would, soon enough, sink into the dense forests and impassable swamps of western Angola, a country still profoundly suspicious of travellers in its isolated reaches. Venture far enough to the north and you would disappear into the tangle of rivers and rainforests of the Democratic Republic of the Congo. In fact, if you were to travel north from Liuwa, in thousands of kilometres you might not cross a single paved road or encounter a town of any substantial size before reaching the Mediterranean.

When I first made plans to visit Liuwa, it was a rough two-day journey from the south, from the Zambia–Namibia border, which is itself a long way from anywhere. Beginning in Namibia's Caprivi Strip at the little-used border crossing north of Katima Mulilo, dirt tracks—barely passable after rains—shadowed the Zambezi River for a time, then meandered through the badlands of the Barotse people who call Liuwa home. Whenever the trail arrived at a riverbank, a decrepit barge would ferry vehicles across, one at a time.

But I took longer to get there than planned and, by 2016, a few months before my arrival and with October rains imminent, a paved road connected Liuwa to the outside world for the first time. The road was, at the time of my visit, yet to be marked on maps and more donkey carts than cars passed along its tarmac. Still, it felt like cheating.

Beyond the Barotse capital of Mongu, a busy town still beyond the pay grade of credit cards and banks willing to take dollars for Zambian kwacha, the road traversed the flood plains that dictate life here for the Barotse people. In good years, the rains sweep down upon the land with monsoonal force in October or November, bringing fish and floods, and prompting a migration of people to higher ground—including, with great ceremony, the Barotse Royal Establishment.

The new paved road ends at Kalabo, on the banks of the Luanginga River, and there I met Jakob Tembo. He was just as I'd imagined him, a towering, square-shouldered man in khaki military fatigues, an AK-47 by his side. Closely cropped hair crowned a head that described a perfect oval, and the merest hint of a beard played across his chin. His eyes bored into me, shifting only to scan the riverbank from time to time. He was sizing me up, I was sure, but his eyes betrayed nothing; whatever his opinion, he kept it to himself. Unsmiling and without malice, he took my hand and nearly crushed it.

'Mr Liuwa' is how the park authorities introduced him to me.

'Call me Tembo,' the man himself replied. Tembo means 'elephant' in numerous African languages, and Jakob Tembo was a formidable enough character to wear such a formidable name without need of any other.

I have known many African soldiers in my time. From Niger to Côte d'Ivoire, in Abidjan and Johannesburg, military men—many smelling of alcohol, others red-eyed from ganja—have shaken me down. On those days when I resisted, long waits in hot sunshine ensued until they grew bored or I yielded, conscious that a busload of passengers could not cross the despot's chain without me. Once, in the pestilential humidity close to Douala in Cameroon, one grabbed me by the shirt and threatened to 'disappear' me. On another occasion, while I waited in line to cross into Burkina Faso, a border guard pointed at my yellow vaccination booklet, not my passport, and told me that my visa to his country had expired and he would have no option but to cart me off to prison. And no matter how poorly I was treated, my experiences were nothing when compared to the indignities visited upon ordinary Africans by these ragtag militias masquerading as national armies. At checkpoints across the continent, I have watched suitcases emptied onto the roadside, bicycles 'confiscated' 50 kilometres from the nearest town, and elderly

African women of great dignity pushed to the ground by these bullies with guns. In the presence of the uniformed men of Africa, I tense my jaw, never quite ready for the unpleasantness that will surely follow.

Tembo was different. He didn't swagger, there was no menace in him, and he carried himself with calm authority. I liked him instantly.

Liuwa Plain National Park lay beyond the river, and we waited in line for a pontoon ferry, an uncertain contraption of cables, winches and no great speed. In time, we took our turn with bicycles, goats, and women balancing baskets on their heads, children bound tightly to their backs. On the far river-bank, we left them behind to forge north, through deep sand, into the Barotse heartland.

As Tembo and I drove on into the wild, I felt the familiar rush of joy that comes every time I enter the parks and reserves of Africa. It is like passing through some magical portal into Alice's Wonderland, the hidden world of Narnia or the inno-cent pleasures high above the Faraway Tree, away from prosaic daily life and into a place where scarcely imaginable creatures live and unimaginable things might happen.

Liuwa evoked a vast inland sea. Golden grasslands, in places as high as the vehicle's windows, bent and swayed in a light breeze. A trail, barely discernible but known to Tembo, parted the grass that sprang back to attention behind us as if we had never passed. An oribi—a small, graceful antelope found on flood plains from Senegal to South Africa—grazed shyly, its head still. Like so many African species whose vast ranges make them appear more numerous than they are, oribi populations are increasingly fragmented across this range and, unusually, this one was alone. These remarkable little antelopes often live in small harems, keeping close, cleverly, to zebras and wildebeest that provide more appetising meals for predators, and can live entirely without drinking, taking all the moisture they need from nutritious grasses. When not feeding, the male oribi spends much of his day spraying from a proliferation of scent glands to mark his territory, but he will break off from this important task when he spies a female urinating or defecating. Unable to conceal his excitement, the aroused male rushes to her, sniffs her hindquarters, paws the ground and then he defecates and urinates on the same patch of ground.

Sadly, the oribi gave no sign of this courtship ritual. Nor did the oribi run as we passed, which, too, was a pity: when an oribi runs, in zigzag fashion, it can reach speeds of 50 kilometres per hour, and is known to leap a metre and a half off the ground to confuse a chasing predator and survey the surrounding scene. This one watched us pass, more inquisitive than concerned.

Where the grasses were shorter, a few wildebeest bulls grazed. The great herds had migrated north and those that remained waited here for their return and for the rains. Immobile, the bulls stood like statues cast in granite, eternal sentinels on the plain, gatekeepers to some lost civilisation. They seemed faintly absurd—their handlebar horns, fly-blown beards and long, eccentric-uncle faces brought to mind the widely held African legend that the wildebeest was cobbled together from the spare parts of other animals left over in the last days of God's creation. Even so, there was nobility in their resolute stance.

Further into the park, grasslands yielded to open plains, vast and without mystery. From sandy soils grew short, stubby vegetation. It was like crossing an ocean floor from which the tide had temporarily retreated; in the distance, trees clung to the horizon and looked like islands. In the shallows of mud-walled pools, spur-winged geese—the largest geese in the world—waded, alert for bugs. Rather unlovely brown birds whose wings flash white when they take flight, their bodies carry the mark of evolutionary genius: they absorb the toxins of the beetles they eat, causing them no harm but killing any other creature (or human) who eats the goose. Take that!

At times the trail was easy, and we made good time. More often, soft sand and deep holes slowed our progress, and as the vehicle lurched into deep and sudden ditches, I learned to listen to Tembo's counsel; he knew every pothole from memory. He talked of Liuwa and her lions, hinting at larger stories. He pointed to one of the islands—Matamanene—where the story of Liuwa and its lions was, he said, unfolding as we spoke. We would go there tomorrow, he assured me, as I made to turn the vehicle in that direction. 'You must be patient.'

I was not in the least patient, and Tembo knew it. I eyed his gun and briefly considered seizing power, but quickly thought better of it. Saying nothing, he directed me towards another of the islands, a campsite called Katoyana, close to the geographical centre of the park. Almost casually, although I already suspected that nothing was casual with Tembo, he warned against straying into the long grass: a black mamba known to inhabit the campsite had been last seen close to here. I fear the black mamba more than any other African snake; it

152

is responsible for dozens of deaths across the continent every year and is, rightly or wrongly, feared for its aggression and for the legends that swirl around the species. The colour of the black mamba is best described as pale silver; it is called 'black' because the inside of its mouth is black. In other words, you will know the snake's true nature only once it is too late, in the split second before it strikes.

Night fell. In the glorious silence of a night free from engine noise, beneath a spray of stars, it was almost unnaturally calm, save for the occasional scrape of spoon on aluminium can as we prepared a simple meal of pasta and tinned sardines. But we were soon driven under canvas by a storm that swept in with tsunamic force, its fire and fury lighting the plains as if it was midday. The local Barotse believe that thunder is the spur-winged goose beating its wings against the land; lightning comes when the poisonous spur on its wings strikes the earth.

It was thrilling, elemental, but if these rains continued my journey would be over before it began.

~

When the Zambian government decided that perhaps it should do something about Liuwa after all, not only was there very little left to save but the government still had no resources with which to do it. So it did what all governments do when they can't do something themselves: in 2002, it put the management of Liuwa Plain National Park out to tender.

It's a difficult enough task to sell a national park project if the park in question is a lawless place stripped of its wildlife.

But in Liuwa, some 10,000 people lived, and still live, within the park's boundaries, complicating an already complicated conservation picture.

Relocating these people was not an option. Ever since the Barotse can remember—and at least since the eighteenth century, when their ancestors migrated here from the south—Liuwa has been a royal hunting reserve, a place where the Barotse Royal Establishment has come for pleasure hunts. Lions were an essential part of this—every year, amid much ceremony, teams of paddlers from specially chosen families would propel the Barotse royal family on royal barges between their summer and winter homes; most of these paddlers wore the mane of a lion. Those who lived in Liuwa did so on behalf of the Barotse king. They were wildlife rangers, centuries before the term existed to describe men like Tembo, and the villagers protected the land from intruders and guarded lions that effectively belonged to the king.

Such was the state of affairs when missionary David Livingstone passed through Barotseland in 1851, and in 1878 when a Barotse king named Lewanika attended the coronation of King Edward VII in London. As was reported at the time, 'He looked a fine and imposing fellow in Westminster Abbey'; one can only imagine the stir he caused as he entered the hallowed cathedral and took his seat. Even so, for centuries the outside world rarely encroached upon Liuwa, and its lion and human inhabitants. A stray hunter, a missionary or two, a sweaty colonial official in pith helmet and khaki shorts—only occasionally did the world beyond make its presence known here. Liuwa was a backwater, even as the country transitioned from

British-controlled Northern Rhodesia (as Zambia was then known) to independence as the Republic of Zambia in 1964.

In 1972 the still-new government, keen to impress Western donors and stamp its often tenuous authority upon distant and sometimes restive provinces, announced that it was taking over the park. The Barotse royals watched on, no doubt bemused, as the government declared a game reserve, then upgraded Liuwa to the status of national park. Whatever the government called it, they knew better than to evict the Barotse custodians who had lived in Liuwa for centuries.

We'll look after the animals from here on in, they told the local people.

'It didn't work too well,' Tembo told me in Liuwa, nearly 50 years later.

The government takeover began Liuwa's long and inexorable decline. What started as perhaps well-intentioned government stewardship of an important wildlife-rich ecosystem descended into something else altogether. Infrastructure was basic, the government paid its rangers poorly, if at all, and these rangers lacked the resources to properly patrol park boundaries. Wildlife survived as much by accident as by good management, and the government was more concerned with its own survival than with protecting a few animals that nobody ever saw. When the government walked away from Liuwa in the 1990s, the situation looked terminal.

So when the government—perhaps feeling guilty, more likely at the urging of a Barotse royal family hinting at secession—put Liuwa out to tender, it was surprising that anyone wanted the job.

Undeterred, two interested parties applied. One was a local community association with widespread support across the region. The other was African Parks, an international NGO that had done extraordinary work in bringing broken national parks back to life across the continent. It was a familiar contest in the new Africa: a local group with strong local connections but few resources pitted against a well-resourced Western organisation with access to deep-pocketed donors.

After a bitter campaign, the Barotse Royal Establishment chose African Parks, and the international community breathed a sigh of relief. But many locals were unhappy, and that discontent simmers to this day.

The decision brought to the fore one of Liuwa's defining narratives: where you come from matters. When a baby is born among the Barotse, the umbilical cord—sometimes the placenta, too—is buried alongside the place of birth. It is a profound symbol of a person's—a people's—connection to the land. From birth, part of them resides in the soil. They are the land. The land is them. If you come from elsewhere, as African Parks did, if your own umbilical cord is buried in a different place, there is the assumption that you are only tarrying here, and that you will return to the soil of your own ancestors upon completion of your business. The land in Liuwa is not yours. You will never truly belong.

~

No one knows how Lady Liuwa came to be alone or how long she endured without company, although people would later try to backfill the gaps with speculation. Some said that Lady had always been a loner or an outcast, even when there were more lions in Liuwa. Others assured me that she had belonged to large pride. But no one really knew.

Based on the accounts of local hunters and on the widely held belief that by 2001 no lions remained in Liuwa, this much can be claimed: by the time that Tembo and Lady became friends in 2004, Lady had most likely not seen another lion in three years, probably more.

This is remarkable for many reasons. For one thing, surviving alone is, for a lion, an astonishing feat. Being a lone huntress for a time is not necessarily an impediment to survival; one study of Serengeti lions found that lone lions fared just as well when it came to food intake as those in hunting groups of five to seven lions. Some lions can do this on their own for a while; nursing mothers, for example, often leave a pride for weeks on end to care for their newborn cubs and must sometimes hunt alone for the duration. As we've seen with the lions of Hwange National Park in Zimbabwe, male lions, too, sometimes spend months alone, searching for their own pride or coalition or territory. But to do so consistently for three years places a remarkable strain upon a lone lion, especially in an open habitat like Liuwa's, and is a feat rarely documented in the world of lions.

Lions must kill regularly—at least every few days, on average—to survive. They most often hunt as part of a pride and prefer to do so, using elaborate tactics like ambushes, decoys, flanking, encircling and running in shifts. Working as a team, lions are also able to down much larger prey than would ordinarily be possible for a single lion, including elephants, which weigh 3.5 tonnes or more. Even adult buffalo can weigh nearly 900 kilograms, around seven times the body weight of an adult lioness. An adult wildebeest, which Lady killed often, can weigh 280 kilograms, more than twice her weight. Although lions have at their disposal certain weapons—raking claws, dagger-like teeth—bringing down one of these large animals can be the human equivalent of wrestling a grand piano or a horse to the ground with your bare hands while it struggles and fights back with sharp horns and deadly hoofs.

Lady's hunting prospects were further challenged by conditions unique to Liuwa. The park is an open land, one better suited to cheetahs who require long sightlines and open country to chase down their prey. Lions prefer cover and rely on getting close to their prey—preferably within 10 metres—without being seen before launching their attack. Islands of light woodland and plains of long grass do exist in Liuwa, but the park's prey species avoid them where possible; a wildebeest's true habitat is the short-grass plains. Even if a lion can get close enough, hunting wildebeest isn't easy: they are as alert as any animal whose life could end at any moment; they seek safety in numbers; and, with eyes stuck to the sides of their faces, they might as well have eyes in the backs of their heads. That being the case, it would take a hunter of exceptional skill to get close enough to strike. Clearly Lady was that.

Having killed their prey, lions together protect their meals from competitors. A buffalo, an adult wildebeest or an eland provides enough food for a week. A smaller kill—an oribi or a warthog, for example—might sustain a lioness for a day or two.

Lions routinely eat immediately upon killing—lions can eat up to one-quarter of their own body weight in a single sitting, which is like an average woman or man eating a nearly-18-kilogram or 21.25-kilogram steak, respectively—but return often to eat over the days that follow. Not Lady. When it comes to protecting meals out on the plains, where everyone is hungry, the maths is simple: a group of hyenas is unlikely to drive a pride of lions from a carcass, but they can easily drive away a single lioness, even one as feared as Lady. With few places to hide the carcasses, and with every hyena equipped with a highly developed sense of smell, it would be almost impossible for her to defend her own kills for long. Tembo once watched her drag a carcass up to 2 kilometres to cover, a test of strength rarely required of lions and one particularly difficult to accomplish after the exertions of a hunt. Even so, Lady would rarely be able to spend a week feasting on a wildebeest at her leisure.

Scavenging was also unlikely to be an option for Lady. Although the popular perception is that lions are the hunters and hyenas the scavengers, the reverse is just as often true. George Schaller's classic study of Serengeti lions found that lions in woodland areas scavenged only 16 per cent of their meals. But out on the plains, like those in Liuwa, lions scavenged nearly half of their meals; most of these they stole from hyenas.

When it comes to popular perceptions of hyenas, *The Lion King*, in which hyenas are portrayed as sinister figures of evil, has a lot to answer for. Hyenas are skilled hunters, charismatic personalities on the African plains and one of the most successful large carnivores on earth—they're often the last large carnivores to hang on when an ecosystem goes to hell. They can even be quite lovable once you get to know them. The only widely held perception that rings true is that lions and hyenas don't really get on, and the spotted hyenas of Liuwa were certainly Lady's nemeses. The spotted hyena population ran into the hundreds and it is highly unlikely that a single lioness could steal kills often from the large clans, each one of which had close to 50 members.

With so much working against her, Lady's story is a heroic tale of survival. Any estimates are certainly more illustrative than precise, but let's assume that Lady needed to kill every three days in order to survive. That means that over the course of three years, Lady was able to approach a wildebeest or other prey across open country, undetected, and then successfully bring her prey down close to 350 times.

Perhaps science offers a partial explanation. There is growing evidence to suggest that being a social carnivore gives lions an advantage: living in families has made them the most intelligent of all large cats. We know this thanks to Dr Natalia Borrego and colleagues from the University of Miami, who set up a test for captive spotted hyenas, lions, leopards and tigers. A piece of meat was hidden inside a box that could only be opened if the animal pulled the rope to release the latch. Lions and hyenas quickly worked it out and clearly outperformed the non-social carnivores—the leopards and tigers—lending credence to what's called the 'domain-general social intelligence hypothesis'. This hypothesis argues that living in social groups makes animals such as dolphins, primates and lions smarter, and not just in a social setting. Being a social creature gives them the tools to innovate when confronted with a problem of any sort.

When it comes to lions, the evidence from their daily lives is pretty compelling: they are clever enough to develop guerrilla group-hunting strategies worthy of military tacticians, defend territories amid shifting alliances in complicated games of strategy and raise their offspring in collective, human-like creches. In Lady's case, perhaps she survived because she was just smart.

But when talking about Lady, there is always something else at play, and here science has fewer answers. Another reason Lady's survival is extraordinary is her psychological state. Female lions live enveloped in a wise sisterhood of grandmothers, mothers, aunts, sisters and daughters that sustains them down through generations; in the animal kingdom, only female elephants share a comparable way of life. Male lions are conflicted, riddled with contradictions: as we've already seen in Hwange, they are essential to the pride, yet any given male is dispensable; they are the king of beasts, yet always watching nervously over their shoulders lest a rival dethrone them; and they are the protectors of the prides, but to announce their takeover they will kill every cub that isn't theirs. Males spend much of their lives alone or in small brotherhoods, dipping into family life when and where it suits them. But female lions thrive on their family ties and are rarely alone for long.

It is always dangerous, and usually without any scientific basis, to project human characteristics onto any animal, but perhaps the only way we can understand Lady's reaction to being alone, her drawing near to Tembo, is to characterise it as loneliness. Think of Lonesome George, the Pinta Island tortoise of the Galápagos, last of his subspecies during the final years before his death in 2012 and for a time considered the rarest creature on earth. It suits our relationship with wild animals, makes them understandable, to imagine that they are like us.

Scientists, quite rightly, explain the social ties of lions in utilitarian terms: lions form prides to help with raising and protecting their young and in defending themselves against competitors. There is ample evidence to suggest that lions, and all other creatures, see the world in similarly dispassionate terms—how else, for example, could nursing mothers mate with marauding male lions who have just killed their offspring? Yes, they will sometimes defend those offspring to the death, but in the natural world, instinct almost always trumps sentiment.

Do animals mourn their losses? Can they experience an emotion akin to sadness? Some of my scientist friends will tell me that this can be, when overstated, unscientific nonsense, the Disneyfication of our relationship to the natural world. I suspect that they are right.

But still I wonder—and it is wonder that draws me to immerse myself in the natural world and its creatures. To watch a pride's females interacting is to understand the deep ties that bind them. They greet each other regularly, rubbing faces, lying in close proximity to one another, playing in a constant cycle of reaffirming these bonds. They seem to be enjoying themselves, to welcome each other's company.

And there's this: lions roar only as a means of communicating with other lions, to warn competitors and to call the pride together. The scientific explanation is that their entire DNA is geared towards passing their days in company—it is rooted in their need to survive. But Lady knew how to survive on her own. So why would she call, if not in the hope that one day another lion might answer?

~

Having attached herself to Tembo and Herbert Brauer in 2004, Lady set about trying to win over her newfound friends in the months that followed.

One time, the two men heard Lady making a kill nearby. No sooner had they set off to investigate than she appeared on the trail they would have to take to reach the kill site. When I visited Liuwa 12 years later, Tembo told me what happened next:

She just stood at the middle of the road. Then we drove closer, about five metres. She turned. She was leading us. We followed her. Immediately, there was a wildebeest, and when he saw the lioness, he tried to stand up. She had wounded it but didn't kill it. She had cut its tendons so it could not move. Now she started playing with the wildebeest. She was chasing, pretending to hunt, acting how she acts when she catches animals. It was as if she knew that she was being filmed. Finally, she jumped on the wildebeest. She pulled it down. Then, while looking at the camera, looking at us, she went to lie down. It was like she was saying, 'That's how I do it.' She was playing with us.

She came often to their camp and would lie down and display in front of the two men, or lie in silence nearby, moving when they moved, sleeping when they slept.

Tembo is the most unsentimental of characters. Although he told me of many local traditions or superstitions without comment, he would, if I asked him, admit that he didn't believe any of them. And yet, whatever Lady's intentions, Tembo reached a strange conclusion: 'She is my friend,' he told me. 'Definitely she is my friend. Whenever I tried to call—"Come on, Lady!"—she responded.'

The surrounding bush fell silent. Tembo held my gaze, talking in a voice just above a whisper as he explained how the scientists whom he told were sceptical and warned him not to get too close to this lioness. And he proved them wrong, calling to Lady in front of them, and they watched her come quietly and lie down close by.

'I have that proof that if I call her, she can come. That's why they are saying that, no, this lion is not a lion, maybe it's some kind of spiritual lion. Even today if she comes, I will call her, and I think she is going to respond.'

'Could you call her for me?' I asked. That I would see Lady while in Liuwa seemed unlikely, and to find her in such a vast space would be difficult. I was grateful for any help I could get.

At this, Tembo stood, faced out into the bush and began to call: 'Hello Lady! Hello Lady! Hello Lady! Come on!' There was silence all around. Lady remained hidden. But there was something in that moment that haunts me to this day—a heightened tension, a sense of the plains suddenly alert to Tembo's half-remembered call.

Without waiting for a response, Tembo sat down and again fixed his eyes upon me.

'When was the last time you saw her?' I asked, unnerved.

'I last saw her in 2015.'

'You haven't seen her for a year?'

'I haven't been in the bush for some time. So we are missing each other. If you want to see her, you just call her. She will come. If you have got the belief that you are going to see Lady tomorrow, you can see her. But if you have doubt, you cannot see her.'

Of course I doubted, although I didn't dare tell Tembo. At the same time, I felt the mystery and magic at large, out upon the plains of Liuwa.

Tembo's eyes burrowed into me. 'You are going to see each other. Definitely.'

~

For the four years that Herbert Brauer filmed in Liuwa, from 2004 until around 2007, Lady, Brauer and Tembo became fixtures in each other's lives. The bond between them grew over time—Lady had not seen another lion in four, five, then six, seven years, and Tembo and Brauer were her constant companions.

But something nagged at the two men. A friendship between human beings and a wild lion is not the natural order of things. As the years went by, they became concerned. Days would pass in camp when Lady did not leave their side, not even to hunt. If it has eaten well, a lion may not need to hunt for a day or two. On one occasion, Lady stayed with them for a week without eating. She followed them around camp, scratching trees to sharpen her claws, keeping within the acceptable distance that had come to exist between them—she never came closer than 3 metres, and they never encouraged her to do so, but nor would she lie down much further away than that. That she could have easily killed either man was no longer part of their thinking. Tembo realised that there were times when Lady needed company more than she needed food, and he wondered whether things could continue as they were.

The park authorities had begun to ask the same questions, although for different reasons. If Liuwa Plain National Park were ever to again grow, if its wildlife populations were ever to recover, lions were an essential piece of the puzzle.

Lions, as apex predators, are the cornerstone of so many African ecosystems. By keeping everything in balance, apex predators can become what scientists call 'biodiversity regulators'. For example, after wolves were wiped out from Yellowstone National Park in the US in the 1920s, the elk population grew out of control. When wolves were reintroduced after an absence of nearly 70 years, the consequences cascaded through the park: elk numbers fell from more than 15,000 to a more manageable 6000, and the once-overgrazed trees such as aspen, willow and cottonwood recovered. Everything from bears to songbirds flourished again.

Scientists are even starting to make the argument that in Africa, where many parks and reserves are profoundly underfunded, saving the lion as a species has become indistinguishable from saving all of wild Africa and its protected areas.

The cost of doing both is the same. Save one and you save the other; lose one and you lose the other. Follow this argument to its logical conclusion and lions become landscape guardians, contributing to the health and survival of many other species that inhabit these 'lionscapes' (as some in the conservation community have begun to call them) into the bargain. In this way, lions can be the final bulwark against oblivion, an ecosystem's last stand against extinction on a massive and irreversible scale.

So, too, for Liuwa. In African savannah ecosystems, lions help to maintain biodiversity balance by killing off weak and injured prey animals and thereby ensuring healthy populations of wildebeest and buffaloes. With prey populations under control, grasslands and all manner of plant species have protection against overgrazing. Keeping a lid on hyena activity, too, would make sure that other scavengers, such as vultures, could return; while in Liuwa, I saw not a single vulture nor a single carcass. Hyenas, with their uniquely powerful jaws, clear up everything, eating even the bones, and often the only evidence that a death had taken place even 24 hours before was the white, calcium-fed, bone-rich dung of the hyena. Bring back lions, and the ecosystem can start to work as it is supposed to.

From the outside, to the untrained eye, ecosystems can appear to be doing just fine. But when African Parks took over Liuwa's administration in 2003, the park was anything but fine. Stripped of its wildlife, overfished, overgrown and utterly unprotected, it had turned a corner towards becoming uninhabitable for both wildlife *and* people. African Parks stopped the decline. Their armed anti-poaching units proved effective, and their campaign to win the hearts and minds of the park's inhabitants slowly rebuilt age-old relationships between the villagers and the wildlife with whom they shared the land. Populations of other keystone species stabilised. Then, African Parks decided, it was time to bring lions back to Liuwa.

But you can't just 'bring in' lions. Where do you bring them from? How do you transport them? How do you make sure they stay once they have arrived? And how do you convince the local people that more lions living in the area is a good thing? These questions and others made the entire process fraught with logistical and public relations peril. African Parks had no choice but to try.

In the dry season of 2008, African Parks began tracking a young male lion in Kafue National Park, 450 kilometres east of Liuwa. In the prime of life, wandering in search of a pride to call his own, he was the perfect mate for Lady. They darted him, loaded him into a 4WD and drove him for eleven long hours along dirt roads to Liuwa. Anaesthesia for animals doesn't last that long, and the lion had to be sedated multiple times lest he wake to find himself lurching over rutted roads in the back of a vehicle. If he woke, the consequences didn't bear thinking about.

When finally in Liuwa, the male lion was laid down in a *boma* in the woodlands of Matamanene. Alongside him they placed a large chunk of meat. The plan was that he would wake, perhaps a little groggy but ready to answer the hunger pangs that would surely groan in his belly. They waited.

All the while, a monumental shift in the dynamics of Liuwa seemed imminent. The new male was important for the ecosystem and the future of Liuwa, and no one doubted the scientific value of what was happening. After two decades of decline, nearly a decade after the lion population had been reduced to a statistical zero, everyone involved knew that this could be a turning point.

But it was impossible not to think of Lady: at a human level, everyone was excited. If all went well, Lady would no longer be alone. Within hours, Lady would be hearing the call of her own kind for the first time in seven years, perhaps longer. Had she already smelled his presence? How would she react when she heard him roar? How long would it take for her to appear?

The team never found out. Upon waking, the male lion—disoriented, his system flushed with the residue of numerous sedatives—panicked and threw himself against the fence. The vet in attendance, Dr Ian Parsons, sedated the male again. But it was all too much for the young lion: he choked on a piece of regurgitated meat from his last meal and, in great distress, died.

No one knows if Lady ever became aware that another lion had just flickered across her horizon; the African Parks team, aware the lion's body was compromised by the chemical sedatives, quickly cremated the carcass of the male lion to avoid poisoning any scavengers who would have come to eat. Even so, we are left to wonder whether she sensed something on the wind. And if she was aware of a new lion presence, perhaps it stirred half-forgotten longings, even hope.

Either way, this was a serious setback for those who hoped to rebuild Liuwa Plain National Park, lion by lion. But it went beyond that: everyone present felt immense sadness for Lady.

~

There was too much at stake not to try again, and eight months after the traumatic death of the male lion from Kafue, the team made another attempt. This time they chose two lions, perhaps brothers. The thinking went that two lions could help each other make the transition, connecting each lion to his old life through an anchor of familiarity. Timing was also important. This time they set out on their mission to Kafue during the wet season. That way, with much of Liuwa underwater, they could complete most of the journey by fast boat instead of poor roads, thereby reducing the hours the lions spent under sedation. During the wet season, too, Matamanene was an island; should the lions survive the translocation, so the thinking went, they would be unable to hotfoot it back to Kafue.

The translocation was a success, at least in the beginning. The lions woke and survived the ordeal. They weren't happy. Groggy from the after-effects of the sedatives, disoriented from their sudden appearance in an unfamiliar enclosure surrounded by an electric fence, they threw themselves against the fence. But they remained fighting fit and survived their first test.

The commotion of snarling lions could not go unnoticed. Everyone held their breath.

No one was with Lady when she heard another lion for the first time in nearly a decade. But Tembo was with her not long after, and he described the moment as 'very fantastic'. He watched as she drew near to the *boma*.

'She came. The other lion was somewhere that side, in the *boma*. That's where they were roaring. She was very surprised. She went to the *boma*, going around, around, around. She walked about a hundred metres. She slept there. And the whole night she was there.'

We can also piece together something of the moment from footage of this first encounter that was captured by Herbert Brauer. In it, Lady is alert, wide-eyed in the way of lions encountering strangers. When such encounters occur in the wild with no fence between them, a lion's first instinct is to be wary. Given that such encounters can be fatal—unfamiliar males will often fight each other to the death, and intruding females can also be killed for straying into lands where they are not welcome—lions always assume the worst, ready to fight or flee at a moment's notice.

In the footage, the males become agitated when they catch Lady's scent. They pace up and down like the wild animals in a cage that they were and, upon seeing her, they charge the fence, posturing and growl-roaring. She approaches, crouched low as if hunting, keen-eyed and taut with tension. She comes near but is hesitant. She circles the fence, as if unable to believe what she is seeing. At once nervous and eager to assert her prior rights to the land, she sprays the surrounding trees with urine, scent-marking her territory.

In our human interpretation of events, Lady must have been overwhelmed with excitement, suddenly in the presence of other lions for the first time in at least eight years. I would like to think that the significance of this moment was not lost on her, that she understood that her loneliness was at an end and that her roars would never again go unanswered. In all likelihood, her instincts took hold and the years fell away. She reacted as any lion would upon finding new lions in her realm: she was cautious, intent on self-preservation.

A fence separated them. Even if it hadn't, the males' first instincts would surely be to mate with Lady rather than kill her: as males looking for a pride to call their own, a lioness would be promising to them, not threatening. And so it was that, minute by minute, hour by hour, the situation calmed. With the fence still between them, each grew to accept the other. Lady stayed nearby, leaving her new companions to hunt then returning immediately afterwards.

The plan was to keep the two males in the *boma* for some weeks, even up to two months, to allow them to settle into their new surroundings. But on the fifth night they broke through the perimeter fence. At first, everyone feared that the lions had fled, or perhaps even that they had killed Lady who was, at first nowhere to be seen. In Brauer's footage, tensions run high as their escape is discovered: 'The boys are out! Let's go!'

But they hadn't gone far: the two males were resting nearby, with Lady not far away. As Tembo and Brauer approached in their vehicle, Lady began her usual tactics of rolling playfully, rubbing against trees, at ease in her world. Only this time she wasn't doing it for Tembo. She was showing off to her new friends.

～

An elephant crosses the plains of Amboseli National Park, in southern Kenya; the view extends west towards Tanzania. Every night, elephants leave the park and head north into the Maasai communal lands that encircle it.
ANTHONY HAM

Around 1500 elephants inhabit Amboseli, compared with 600 in the late 1990s. Most feed within the fertile swamps in the park by day. ANTHONY HAM

The cubs of Nempakai and Nolakunte play in the relative safety of Amboseli National Park in October 2011. Life for lions outside the park is far more complicated. ANTHONY HAM

Meiteranga Kamunu Saitoti, the Maasai warrior who killed five lions before becoming a Lion Guardian, on the Eselenkay Group Ranch in the Amboseli Basin in October 2011. ANTHONY HAM

Beaded bracelets and sometimes hand mirrors are an important adornment for the Maasai, including for Mingati, killer of three lions before he became a Lion Guardian. ANTHONY HAM

Mount Kilimanjaro (5895 metres), Africa's highest mountain, rises in Tanzania but is seen here from Kenya's Amboseli National Park, with acacia trees in the foreground. ANTHONY HAM

The view north-west from the heart of Amboseli National Park. Beyond this acacia lie the hills and plains of Maasailand, which extend across southern Kenya to the Masai Mara in the country's far south-west. ANTHONY HAM

Cecil drinking after having just seen off two leonine intruders in 2012. When this photo was taken, Cecil had ruled over Linkwasha in Hwange National Park for nearly three years. BRENT STAPELKAMP

In 2012, Cecil was nearly nine years old, at the peak of his fame, popularity and power, lording it over a pride of more than twenty lions. By this time he had sired eighteen cubs. BRENT STAPELKAMP

Cecil in May 2015. At the time, he ruled over Ngweshla with Jericho, and the pride had four females and seven cubs. He was eleven years and eleven months old. BRENT STAPELKAMP

One of the last photos of Cecil, taken as he walked towards the Hwange National Park boundary and into the hunting concession where, on 2 July 2015, he would be shot by Walter Palmer. BRENT STAPELKAMP

Cecil lying next to Jericho on the last morning that Brent Stapelkamp saw them. Cecil was killed not long after. BRENT STAPELKAMP

Elephants and baboons at Mbiza, an area of Hwange National Park between The Hide and Linkwasha, held by Phobos in late 2019. ANTHONY HAM

The Madison region, between Makalolo and Ngweshla, was occupied in September 2019 by two of Cecil's former 'wives', a four-year-old son of Cecil and four cubs. ANTHONY HAM

The Ngamo pride crosses the railway line that marks the boundary between Hwange National Park and the hunting concessions where many lions have been shot. PAUL FUNSTON

Ngqwele patrols his territory close to Linkwasha Camp in September 2019. At the time, he ruled over both the Linkwasha and Ngamo prides.
ANTHONY HAM

Netsayi, born in December 2014, and his half-brother Humba ruled Cecil's former pride into early 2020. Here Netsayi plays with one of the pride's cubs.
PAUL FUNSTON

Females and cubs of the Ngweshla pride including Cecil's grandchildren wait at a waterhole hoping that a stray elephant calf will come to drink.
PAUL FUNSTON

Ngqwele surveys his territory near Linkwasha Camp in September 2019. As a lone pride male, he was vulnerable to incoming coalitions of male lions.
ANTHONY HAM

Netsayi at Madison in September 2019. Charismatic and beautiful, Netsayi has become one of Hwange's most famous lion celebrities among safari-goers and park guides. ANTHONY HAM

Humba hangs out with one of the cubs of the Ngweshla pride. When not patrolling their territory, Humba and Netsayi are often seen in the company of their cubs. PAUL FUNSTON

In the oppressive heat of October, Humba and members of the Ngweshla pride rest in the shade of young Rhodesian teak trees. PAUL FUNSTON

SPIGF3, one of Cecil's former 'wives', looks out over the Madison waterhole at sunset in September 2019. ANTHONY HAM

One of the Ngweshla females breaks into a run, either to hunt or to chase away vultures from a carcass, around one of the Ngweshla waterholes.
PAUL FUNSTON

One of the cubs of the Ngweshla pride, born to one of Cecil's daughters and fathered by either Humba or Netsayi, plays with a small, rather unhappy leopard tortoise. PAUL FUNSTON

Lady Liuwa in terrible condition in Liuwa Plain National Park, Zambia, in October 2016. Although barely able to walk, she survived for ten months after this photo was taken. ANTHONY HAM

Above: The remote salt pan and grasslands where I last saw Lady Liuwa in October 2016. ANTHONY HAM

Left: The endless horizons of Liuwa Plain National Park in western Zambia. ANTHONY HAM

Below: Two Angola green snakes wrestle at Matamanene in October 2016. ANTHONY HAM

The first storm clouds of the rainy season build over Liuwa Plain National Park in October 2016. By November, the park's plains begin to disappear underwater and most of the park is accessible only by boat, often until April.
ANTHONY HAM

Around 10,000 Barotse people live inside Liuwa Plain National Park and rely on fish caught in seasonal waterholes for much-needed protein. ANTHONY HAM

At the end of 2016, Liuwa Plain National Park was home to 26,000 wildebeest, a number only exceeded in the Serengeti. ANTHONY HAM

Jakob Tembo, 'Mr Liuwa', wildlife ranger for African Parks in Liuwa Plain National Park. ANTHONY HAM

Induna Mundandwe, Barotse chief, at Matamanene in October 2016. ANTHONY HAM

Young San man dressed traditionally as part of a hunting-and-gathering expedition for tourists in the Kalahari in northern Botswana. ANTHONY HAM

Grasslands and salt pan in Khutse Game Reserve in the southern Kalahari, south of the Central Kalahari Game Reserve (CKGR). ANTHONY HAM

The salt pans at Makgadikgadi in the northern Kalahari, part of the largest network of salt pans on earth. ANTHONY HAM

An elderly San hunter re-enacts a traditional hunt for tourists at a luxury camp in Makgadikgadi Pans National Park, northern Botswana. ANTHONY HAM

Dabe Sebitola, San guide, in the Makgadikgadi region. He spent his childhood living a traditional life until the San were expelled from the CKGR in 1997. ANTHONY HAM

A San woman in traditional dress in Makgadikgadi Pans National Park. The San's presence here dates back 200,000 years and they are one of the oldest peoples on Earth. Some of the earliest records of modern human beings have been discovered near here. ANTHONY HAM

Ally Sefu Mlimile, attacked by a lion in the Rufiji River region of southern Tanzania in 2003.
ANTHONY HAM

Mtoro Mohamedi Ngogi, who was still a child when dragged from his bed by a lion in 1998.
ANTHONY HAM

Semeni Nasoro Malenda, mother of eleven-year-old Sadiki Dubuga who was killed by a lion in 2003.
ANTHONY HAM

Shamti 'China' Ngaona and his family were attacked by a lion while they had dinner in 2003.
ANTHONY HAM

A *dungu*, a two-storey shelter of the kind where many farmers slept and were attacked by lions while guarding their fields along the southern bank of the Rufiji River. ANTHONY HAM

Looking across the Rufiji River towards the south bank where many of the lion attacks occurred. The river runs through the Selous Game Reserve in southern Tanzania. ANTHONY HAM

An elderly lioness in the Selous Game Reserve. The Selous is home to one of the most important lion populations in Africa. ANTHONY HAM

Netsayi, part of the ruling coalition at Ngweshla in Hwange National Park, Zimbabwe, and the latest lion celebrity where Cecil once ruled.
PAUL FUNSTON

At first everything went well. The new males mated with Lady, raising hopes among Tembo, Herbert and the African Parks team that a new pride was being created before their eyes.

It also became clear that Lady hadn't forgotten her old friends. On one occasion, not long after the male lions joined Lady, Tembo and Brauer were filming at night, using a spotlight to illuminate their subjects. Filming finished, they tried to restart the car but the spotlight had drained the battery.

I have feared such a moment for as long as I have been travelling alone in the African bush. Turning off the engine is a basic rule of watching wildlife from a vehicle: not only can you watch in silence but you avoid camera shake. Once in Tanzania's Serengeti, I was watching two male lions asleep under a tree from a distance of perhaps 8 metres. When it came time to move off, the engine was dead, its mortal 'click' sickening. Fortunately there were other vehicles around and they parked between the lions and my vehicle, shielding us from view as we tried to get things started. But I travel alone often enough in remote regions that, one day, surely, my luck will run out and I will be faced with the choice of getting out of the vehicle within sight of lions or waiting an eternity until they move off. That was the choice that faced Tembo and Herbert in the rainy season of 2009.

Accustomed as they were to Lady's presence, Tembo and Herbert took turns to get out of the vehicle and push. The reaction of the male lions was almost instant and entirely as you would expect from wild lions: ears pinned back, eyes wide, bodies ready to break into a run, they began moving quickly towards the two men.

'They came directly towards us,' Tembo remembered. 'At first they were about 20 metres away. And then, as they were coming closer, Lady just came straight to us and sat between us and the male lions. She stood there, not allowing the males to come closer.' When they tried to pass Lady, she 'slapped' them. 'Lady was protecting us, not attacking. I have never seen a lion that can defend a human being.' There was still disbelief in Tembo's voice seven years later as he described what happened. Finally, they changed the battery and the vehicle started.

Later that night, with Lady nowhere to be seen, events took a sinister turn.

'Around 4 a.m.'—or 'zero-four' as Tembo says it, military style—'they came. I heard some twigs being broken. I knew that it was the lions. I was just quiet inside the tent. Then I heard one lion coming to where my head was, the other lion where my feet were. I heard them pull down the top layer of the camping tent [the fly]. I was inside. It was then that I knew I was being attacked.'

Let's just pause for a moment to consider what was happening. I have slept under canvas out in the African bush. I love the experience. It is exhilarating, allowing you to feel at one with the night, yet protected. This is usually true—. attacks on tents are rare—but expecting a tent to withstand a lion attack is akin to trying to stop an elephant charge with a handkerchief. Standard-issue tent canvas would be no match for the claws or teeth of a lion. And even if you could exit the tent safely, a pair of male lions would cut you down before you could run 2 metres. Tembo knew that he had seconds in which to save his life.

'I got my gun. It was half-cocked. It wasn't ready to shoot. One of them hit against the tent with his paw, like this'— Tembo raked his clawed hand downwards. 'He made a big hole in the tent. I moved my leg up. He got a shock. He roared. He was afraid. Then I tried to cock the firearm but couldn't. But then I cocked the firearm again and I fired the gun. I was inside but I shot up above the lion. Then they ran away.'

After twenty minutes, with Tembo watching from a short but safe distance away, the two male lions returned, grabbed the tent, tore it to shreds and carried it off into the night. Tembo and Brauer drove off, leaving behind the dead car battery to be picked up later. Before they could do so, the two male lions destroyed the battery.

'They were not happy with us,' Tembo told me. You think?

~

Then things started to go wrong.

It quickly became clear that the mating of the male lions with Lady, witnessed by Tembo and others, had come to nothing: Lady never became pregnant. Either Lady was past her prime or infertile. No one knew if Lady had given birth to cubs in the past. All that mattered was that if a proper pride of lions was to again roam Liuwa, more females would be needed. Lady had held the line. It was now up to others to take the lion population forward.

In 2011, African Parks brought in two young lionesses, sisters, hoping that they would mate with the two males. It was a risk. Although unrelated females occasionally do link

175

up to form a pride—living in the same territory, watching each other's backs, sharing the same food—such instances are rare. Save for unusual cases in the Serengeti and elsewhere, only in the depths of Botswana's Central Kalahari Game Reserve have cases of unrelated females joining together been regularly documented, and there it is almost certainly a dry-season arrangement dictated by arid conditions when prey is scarce.

Initial results were mixed. Perhaps the new lionesses and Lady just didn't get on. Or maybe she had grown too used to being on her own, too set in her ways. In 2012, in an echo of Liuwa's past, a poacher's snare claimed one of the females. A short time later, the other young lioness, Sepo, fled towards the Angolan border. The African Parks team, who were tracking her radio collar, knew that they had to act quickly; once she crossed into Angola where they couldn't follow, she could be gone forever. On the verge of a very different destiny, she was darted, flown back into the park in a helicopter and locked up in the *boma* for two months, this time with Lady for company. It is not on record what Lady thought of such confinement. The two lionesses bonded in the *boma* and, by the time they were released, they were firm friends who went everywhere together. The long years of solitude must have seemed—if Lady had turned her mind to it—like a distant memory.

Things were definitely looking up. Not long after they were re-released into Liuwa, it became clear that Sepo was pregnant; lion pregnancies last for around 110 days, so Sepo and the males must have mated just prior to her confinement. A short time later Sepo disappeared from view, this time for good reason: as female lions always do, she chose a secluded, sheltered area and gave birth to three baby lions, the first cubs to be born on the plains of Liuwa in more than a decade. The dream of a Liuwa pride of lions was starting to come true.

But life in Liuwa is never that simple. Soon after the birth, the two males, roaming as males do, left the park and crossed into Angola before the African Parks team could stop them. There, villagers shot and killed one of the males. The other, the last male standing of the three who had been introduced into Liuwa, hightailed it back to the sanctuary of the park. In 2014, this last male was found dead, either from disease or poisoning. As always in Liuwa's recent history, for every step forward, there seemed to be two or three steps back.

And there was more trouble brewing for Liuwa's lions. Many local villagers—those living inside the park and those around the park's perimeter—were none too happy that lion numbers were again on the rise. African Parks had been careful to consult with as many people as they could, sending out a questionnaire to locals, asking for their thoughts. Some among the locals were in favour; according to Tembo, they said something along the lines of 'Okay, lions can come. It is our culture. Lions have been here for many years'. But reintroducing lions is no abstract thing when you live in communities that lie within lion territories, and for all the talk of rebuilding ecosystems, for all the benefits that lions bring to a land, lions inspire fear: the fear of being killed by a lion lies deep in the human heart. As Tembo well knew, the line between a man-eater and a lucky escape was often canvas-thin. When the results were tallied, nearly half of the local population opposed the return of lions to Liuwa. Despite the opposition, the Barotse Royal Establishment gave their blessing to bring in more lions.

~

The morning after I arrived in Liuwa in October 2016, we drove to Matamanene. En route Tembo filled me in on the latest news. Big things were happening.

Sepo, Lady's female companion, had in the past few days disappeared into a thicket on the outskirts of Matamanene. No one wanted to get close enough to disturb her, but they knew where she was—the beep from her radio collar was constant and stationary. It was Liuwa's worst-kept secret that Sepo had either already given birth to her second litter of cubs, or was doing so as we spoke.

In further exciting news, just the night before, a new male lion had arrived in Liuwa, brought in from Kafue. We were on our way to see him in the *boma*—the same enclosure where the first male had died, where the two males had broken through the fence, and where Sepo and Lady had bonded. In the *boma* with him was Liuwa's resident male, a youngster from Sepo's first litter. In a high-risk strategy, the two sleeping lions had been laid down next to each other the night before. The morning would be the lion equivalent of rolling over and waking up, groggy, to find a stranger alongside you in bed, and wondering at first 'Where am I?' then, 'Oh God, what did I do last night?'

The hope was that they would bond and, upon their release, form a new coalition to rule over Liuwa. But male lions who don't know each other very rarely get on—Cecil and Jericho were exceptions—and will often fight to the death when they meet. Tembo did not yet know what had happened when they had woken, and we raced towards Matamanene, eager to learn more.

The plan was to keep the two males in the fenced enclosure for up to eight weeks, but the danger was that the new lion would kill Sepo's new cubs immediately after being released. Although emails from the African Parks team had been upbeat in advance of my visit, I knew that this was a high-risk gamble not of their own making. They had tried to bring in the new male well before Sepo became pregnant to avoid exactly this scenario, but the wheels of bureaucracy and conservation, and the tricky process of lion translocation, move slowly, and Sepo was already pregnant by the time the paperwork came through.

The stakes could not have been higher for Liuwa's lions. If all went well, Liuwa's five lions—Lady, Sepo, and her three subadult offspring—would become eight. (It was later confirmed that Sepo had given birth to two cubs.) If things didn't go to plan, the new male could kill the resident male and the new cubs, perhaps even killing Sepo in the bargain if she fought to defend her cubs; Liuwa's lion population would then be reduced to four. A pride of eight or a frightened remnant of four: such were the fraught and finely balanced margins of Liuwa's lion future.

Home to African Parks' base inside the park and to the scientific team from the Zambian Carnivore Programme, Matamanene was a hive of activity when we arrived. Everybody was excited: the male lions had passed their first test. When the Liuwa male had woken up to find himself alongside an unknown male from Kafue and the carcass of a wildebeest, 'He was just so chilled,' Rob Reid, the bearded, khaki-clad park manager, told me. 'He woke up, just looked over, a little glance, touched him with his paw, and that was the sum total of the aggression. He just sat down next to the wildebeest.' The *boma*'s perimeter fence had taken a hammering from the Kafue male, short-circuiting the electrics, but the fence was intact and the two lions seemed happy enough in each other's company. So far so good. A bullet dodged, at least for now.

With Dr Ian Parsons, the vet who had overseen every lion translocation here in Liuwa, we approached the *boma* with great caution, careful to remain in our vehicles. Nobody wanted the males to become more agitated than they already were—the old rule about lions reacting differently to people on foot held here, despite the fence between us.

Through the tangle of branches and browning leaves of the late dry season, we could just make out the two lions, as tawny as the vegetation and with the peach-fuzz manes of adolescents. They sat side by side, Sphinx-like, alert and on edge but more concerned with us than with each other. Both would have seen vehicles before, but their stillness surely masked an inner turmoil stirred by the strangeness—the unfamiliarity of the surrounds for the Kafue male, the worrying lack of freedom for both of them, the uncertainty surrounding their new relationship—of what was happening to them.

It was a stolen glance. The lions were difficult to see through the bushes, like some grainy, indistinct image of a nearly extinct animal. But knowing what this meant, I was as excited as a child. We all were.

We withdrew, out of sight of the *boma* but still within earshot. Reid had an easy, laconic way about him. Like so many South Africans, he ended most statements of fact or opinion with a half-questioning rise, an 'eh?'. He also pronounced wildebeest as *voldebest*, as it's supposed to be said. Respectful of the local people, he knew that they grumbled behind his back and suspected him of all manner of treachery; he had even been accused of wanting to kidnap Lady and spirit her away to South Africa in the team's tiny plane.

Back when African Parks took over in 2003, Reid explained, as few as eight or ten wildebeest remained, and other species had either disappeared or had been reduced to a few stragglers scattered far and wide. To supplement the existing population, African Parks brought in hundreds of animals from other parks across the region. By the time we spoke in late 2016, Liuwa was home to at least 26,000 wildebeest, the second-largest population on the planet after Tanzania's Serengeti. An estimated 4000 zebras, too, roamed the plains, although another study put the figure closer to 9000, while the 500 tsessebe—an oddly shaped, uncommonly fast antelope with an elongated head and hindquarters that look as though they've been dipped in a paint pot—now made up one of Africa's healthiest such populations. Spotted hyenas rule Liuwa, with 350 in the park's core area, and more than 500 across the ecosystem.

Reid well understood the importance of lions to Liuwa. If Liuwa was to survive in the long term, it would need to attract enough tourists to make the park self-sustaining. And as everybody knows, in the aftermath of *The Lion King*, it would be impossible to attract anyone to Liuwa with a promise of cackling hyenas.

Although it would be many lion generations before Liuwa's lion population returned to its former size, the interim aim, according to Reid, was far more modest. 'You don't want too many lions, because then the villagers get upset and poison a wildebeest—one poisoned wildebeest could take out the entire pride and all this effort goes to naught. Ideally, we would have two prides. It's finding that balance.'

Reid also knew that Lady was the thread that made it all possible when it came to convincing locals to accept a growing lion population.

'The older generation knows that there were lions here. If you lose out that memory, that institutional memory of lions being in the park, you bring back lions in ten years' time once Lady has gone and it's going to be that much harder.'

~

Life on the road can be exciting. More often, it involves doing very little at all: waiting for an animal to appear; for an animal to do something interesting; for the long, hot hours of midday to pass; for someone I'm supposed to interview to turn up. While I waited at Matamanene in the humid build-up to the rains for one of the local chiefs to appear, we charged our phones and laptops off the generator, and talked about lions. Baboons scampered through the undergrowth. In a vivid splash of purple, a violet-backed starling flew down from the trees to drink from a water-filled hollow. Close to the mess tent, two Angola green snakes writhed in mortal combat, a disturbing dance of death that left us all feeling solemn, their vivid green seeming to dull as blood and exhaustion soaked through them; such moments in the heat of the African day are as common as they are strange.

While we waited, Tembo and I talked. He first arrived in Liuwa in 1991, aged 25; at the time of my visit in 2016 he had been there for a quarter century—half of his life. He married a local Barotse woman and together they have six children, all born upon the soil of Barotseland, all living in Kalabo close to Liuwa.

'I am happy,' Tembo told me, 'but I cannot stay here.' The local people 'are always fighting for land, so how can they accommodate me?'

Tembo knew that he would never be considered a local. It didn't matter that everyone, including a prominent local chief, called him 'Mr Liuwa' in recognition of his knowledge of the land and his long years spent protecting it. Most likely, Tembo knew the plains of Liuwa better than any local. In spite of everything, he still didn't belong. This was not where his umbilical cord lay buried.

Tembo was resigned to his fate.

Only once did I ever see Tembo at his most frightening, and for the briefest of moments he was the protector of Liuwa, sworn enemy of poachers. It happened one day when we were driving back across the plains towards Katoyana campsite, and we happened upon two lone young men on foot. They had no business being in such a place close to sunset. Tembo bristled.

There is nothing romantic about being a wildlife ranger in Africa. Vigilance defines what they do; their own sudden death is always possible. Tembo's transition from our friendly conversation to alert suspicion was instant and uncomfortable. His lip curled, his rapid-fire questions dripped with hostility, and he made a point of visibly placing his hand upon the gun. They asked for water. He refused. He knew they weren't locals. The air was thick with fear and tension. We drove on.

And yet, as we sat in Matamanene, Tembo spoke of how he had planned to purchase a block of land not far away— a farm, east of Mongu. He wanted to put down roots, thought it was time. But what stopped this man-mountain with a gun was the intervention of his own mother: 'My mother would not be happy for me to be buried here, where there are none of my parents.' He knew because she forbade him to buy the farm. 'My mum said, "No, no, no. I don't let you do that." My mother told me, "No, it must be part of Eastern Province, close to where I live and where you are from. Then I will be happy."' Tembo's reply, as he told it to me? 'Okay, thank you, Mother, I will do that.' Deep down he knew that it was futile to argue with his mother and her call to one day return to his homeland, just as it was futile to fight for his right to belong here among the Barotse.

As for people, so too for lions.

While a resistance to lions is entirely normal, even common, across Africa, there was more to local opposition in Liuwa than a fear of lions and the threat they posed to life and livestock. To the Siyenge, the descendants of those who had once been charged by the Barotse king with protecting the lions of Liuwa,

Lady was one of them. Whenever they saw her, they would kneel, bow their heads and clap softly in a sign of respect; prior to my visit, I was taught an identical protocol by the park's authorities for whenever I met a local chief. These Siyenge would call to her, greeting her as she passed, even when she was far away. To them, Lady was a profoundly spiritual creature, and local Siyenge stories, never before told beyond Liuwa, held that Lady was an ancient Barotse ancestor, a respected old woman from the Barotse creation myth who had turned herself into a lion.

If Lady was their last sacred lion, then these new lions—the new male, Sepo, those that had died—were, like Tembo, imposters with whom the Siyenge shared no history. They were unwelcome intruders on sacred lands.

This was confirmed at Matamanene by Teddy Mulenga Mukula, a field ecologist with the Zambian Carnivore Programme and also an outsider; he was from Zambia's north, and had been with the Programme for three years. Well aware of the local cult of Lady Liuwa, he had come to understand that, for the most part, the new lions were not welcome. 'The local people resisted,' he told me, 'because they believed that those lions are not from this place. They believe those lions will destroy their cattle, their goats.'

He drew the analogy with people: 'Even in the same Barotseland, you are not considered local everywhere. No, you should have a place where you buried your umbilical cord.'

And the new lions?

'They were not local lions. That's just a fact, because you are talking of the ancestors who had become lions, their forefathers from families that have originated and have continued in here. And they put names to lions that are born here.'

What about the cubs? Weren't they born upon the soil of Liuwa?

'I have heard it said that when they are born here, like the cubs that we have, they will accept them because they are born here.'

And then a thought occurred to me. Lady may not have had cubs of her own, but by her unbroken presence—by making it possible that lions remained in the memory of local communities—she served as a bridge, enabling a time when lions could once again be born upon the soil of Liuwa. That was her legacy.

True to her remarkable affinity with humans, she went a step further. As Mulenga Mukula told it, 'Lady has done a very tremendous work in terms of keeping the pride in check. It's very rare cases that we have heard that the lions have killed cattle. You would expect that it happens a lot here, where people are inside the park. But it's very, very rare. She has taught the new lions not to attack cattle.'

~

It's not every day that you meet an African chief, and when Rob Reid introduced me to Induna Mundandwe—*induna* means 'chief' in the local language—I half-crouched, half-bowed and clapped softly as is the protocol. I felt self-conscious and more than a little foolish, and the chief chuckled softly, also in embarrassment. I think we were both glad when I finished.

Induna Mundandwe was as responsible as anyone for the return of lions to Liuwa. A shy, jovial man in striped polo shirt, grey windbreaker pants and smart black running shoes, he wore the regalia of his office lightly. Only when pressed did he pull from his pocket his red ceremonial headgear that we both agreed looked like a shower cap.

Here was a reluctant chief, the mantle of leadership having been thrust upon him when his elder brother was passed over for health reasons. Born into a leading local family in 1964, he grew up in Injera village within the boundaries of what is now Liuwa Plain National Park, perhaps 15 or 20 kilometres from where we spoke at Matamanene. Back then, when he was a boy, lions were plentiful, and family members regaled him with stories of lions killing people, among them a cousin of his father. Every morning he would walk to school through lion country, and he was never allowed out of the family compound after dark for fear of lions. His parents would sometimes keep him home from school for days when lions had been heard roaring nearby.

He never imagined that he would become a chief: with older brothers and uncles in abundance, there were others ahead of him in the line of succession. His cousin was chief before him and, with no great weight of expectation upon his shoulders, the future *induna* left Liuwa when still a young man, eager to immerse himself in the political maelstrom of post-independence Zambia. 'I didn't like these traditional things,' he told me. 'I wanted to be very independent.'

With each new stage of life, he left Liuwa a little further behind, finishing secondary school at Livingstone, hundreds of miles to the south-east and close to the Zimbabwe border, before moving on to the University of Zambia in Lusaka. His involvement in student politics brought him nothing but trouble—he failed to finish his studies after a 'confrontation' with the regime—and his onward trajectory through life was swiftly curtailed. At the age of 28 he was back in Liuwa, a frustrated young man with no great direction in life.

Eight years after Induna Mundandwe returned to Liuwa, in 2000, his family approached him to become chief. He was, they said, an educated man and his talents were going to waste. He had little choice but to accept, and, anyway, it wasn't like he had anything better to do.

There are ten *indunas* in Liuwa, and each reports directly to the Barotse royal palace, at once the voice of the people and the representatives of the royal family among their subjects. When it comes to seniority, all *indunas* are, in theory at least, created equal. But Induna Mundandwe's realm stood at the very heart of the national park and his 5000 subjects had more at stake in the future of the park than most; although he wouldn't say so directly, he was the de facto leader of the ten *indunas*. He was also the most worldly among them and kept in touch with the outside world by radio—as a devotee of the Voice of America network, the chief was eager to talk about the US presidential elections when we met.

It had been an important time for him to become chief. The Zambian government, perhaps at the urging of the Barotse Royal Establishment, had only just put the Liuwa Plain National Park restoration out to tender. When Induna Mundandwe sided with African Parks, amid the tension between the powerful outsider NGO and the locals born into the land, the young chief found himself at the epicentre of a bitter struggle. It was, he admitted years later, 'painful' and 'a very unpopular decision'. Some said that African Parks were concealing their true motive, which was to rob the people of the land. The chief, others said, was only voting this way because he was receiving financial inducements from African Parks. 'Whatever bad things are going to happen to us,' people told him, 'we shall cry over your body. You are the one who is going to bring us suffering.' Even some of his family members accused him of selling out his people.

As difficult as this was for him, and for all his quiet conviviality when we met, one suspects that Induna Mundandwe relished the rough-and-tumble of debate and accusation; perhaps it reminded him of those roiling days at the coalface of political agitation as a student. He stood firm.

And yet his stance during the whole episode has pursued him to this day. When African Parks wanted to bring in lions, many among his subjects pointed out that African Parks' rivals, the community association, would never have approved such a thing. From his earliest days as *induna*, he had been aware of Lady. Back then, he recalled, it was just her and 'another male lion, very big and old. Unfortunately, he just disappeared'— probably, in Induna Mundandwe's account, 'poached out'.

But this shared history meant nothing to locals. If the chief hadn't chosen the foreigners, they, the local people, the real inhabitants of this land, wouldn't have to run the gauntlet of lions and other such nonsense. And Induna Mundandwe was not deaf to their concerns. 'Frankly speaking, people fear lions,' he told me. 'The generation that was used to lions is slowly dying out. It's the new generation that has been born without lions who are not welcoming these lions easily.' He thought about it for a moment, then concluded: 'People tend to resist things that are new.'

For a certain section of the community, Lady didn't count. Or, perhaps more accurately, she was the only one who counted. Confirming what Teddy Mulenga Mukula had told me, the chief acknowledged that for many in Liuwa, the new lions were 'foreign lions'. These people felt 'detached from them spiritually'.

Now the worldly Induna Mundandwe was not one to believe in all of the magic and tradition that swirls around Lady, Liuwa and its lions. And yet he, too, had experienced Lady's affections—a night when she lay down and began purring right by his campsite; when the chief was hosting the Barotse queen at Matamanene and Lady led Induna Mundandwe to his tent, then stepped aside to let him pass; when she appeared in his dreams, and he woke to find her watching him. 'There are not many lions that behave like that,' he laughed, a little self-conscious.

Warming to his tale, he spoke of his own subjects, the Siyenge, who believed that when an elder among them died, that elder became a lion.

And then, just as it had when Tembo had tried to summon Lady to our campfire, the mood suddenly changed. I have no other way to describe it other than this: the earth fell silent. Both of us sensed it. We shifted uncomfortably. I looked around us nervously.

'I remember when I was young,' Induna Mundandwe began—he spoke in barely a whisper but I started at the shock of it—'that when an elder of the village died, the lions would come to the village and start roaring, and they would continue roaring throughout the whole night. Sometimes the lions would kill an animal, like a wildebeest, pull it to the centre of the village and leave it there for the mourners. There was that strong belief that it was their ancestors who had done that, that they had come to mourn one of their own.'

And then he told me the story of Lady's true identity. 'She is Mambeti. This woman was an old woman who lived a very long life. She was a very old woman who took care of the clan. And she was very kind, so she was given a lot of respect. And when she passed on, it happened that the lions did what happens when an elder dies: they came roaring just like that. So the elders, seeing that most of the lions were killed, that it was only her, this lioness who remained, they said, this must be Mambeti, the last to remain from her generation.'

The whirr of cicadas resumed. The flies returned. Induna Mundandwe paused, this modern man ruling over a traditional people, himself like Lady a bridge between past and present, holding the centre lest it break.

'I am a product of the twentieth century,' he continued, 'so sometimes I become sceptical about these things. To me she is just a fortunate animal that survived. But then sometimes you are forced to believe, when you see something like this with your own eyes.'

~

I had heard so much about this extraordinary lioness that it was indeed time to meet Lady, to see her with my own eyes.

While I had been speaking with Induna Mundandwe, Rob Reid had taken off in the park's tiny Piper plane, hoping to get a fix on Lady's presence somewhere out there in the vastness of the park. As my conversation with the chief drew to a close, Reid returned with the news that Lady was asleep under a palm tree in the remote northern reaches of the park; she had retreated far from the clamour and excitement. With Tembo to guide me and Induna Mundandwe eager to come along, we started out in the mid-afternoon to find her before she set off on a night of hunting. I knew that it was my last chance to see her.

North of Matamanene, the tall grasses retreated momentarily, and we crossed plains so vast and so empty that the sky curved noticeably into the earth at the horizon's outer rim. Wildebeest stood singly or in straggling groups, caught between the migrating herds coming from the north and the old bulls that had remained in the south. Storm clouds billowed, haloed by the late-afternoon sun. The islands here, smaller than those in Liuwa's south, rose as palm trees, dwarfed by forge-black, anvil-shaped clouds and by the plains that turned again to grasslands, golden wherever touched by the sun. Away to the west, ominous fires rose out of Angola and filled the western sky.

The prospect of seeing Lady cheered us all. In the back, Induna Mundandwe, that reluctant leader, spoke of his dreams for his park and his people, of a time when Liuwa Plain, rich in lions and other wildlife, might become a rival to the Serengeti.

Alongside me, Tembo told stories of Lady, of how although Lady could never have cubs of her own, she had become a loving adoptive aunt, babysitting Sepo's first cubs when they were young.

At a signal from Tembo, we cast out from the safety of the track and crawled across a stubbled plain, that endless ocean floor of Liuwa waiting for rainy-season waters. Soon there were grasses as high as the windows and Tembo climbed onto the roof so that he could guide me; a sleeping lioness would not be visible in such conditions and running over Lady was not how I wanted this story to end.

Following Reid's directions and using Tembo's bush smarts, we meandered across the plains, looking for paths where the grasses parted, skirting salt pans lest we sink beneath the surface crust, all the while drawing near to two raffia palms that rose upon the horizon. An oribi broke cover, startling me, and my foot touched the brakes. As I did so, Tembo's AK-47, which he had left in the car standing barrel up on the passenger side floor, moved slowly in my direction until it pointed at my head; I eased it back into its upright position.

We inched closer, straining for a glimpse of Lady. Tembo adjusted our direction by degrees. 'Slowly . . . slowly . . . to the right a little . . . slowly . . . stop.'

Just beyond tall reeds that encircled an open pan lay Lady, this famous old lioness. Two eyes, liquid amber, watched us. She was the colour of the grasses—tawny, golden, the hue of savannah Africa—and all we could see were the eyes, the twitch of an ear, the flick of her tail. Whether disturbed by our presence or because this was the time when lions rise from their daytime rest to begin a night of wandering and hunting, Lady rose and began to move across the pan. We circled away to the east, then south, skirting the pan and its treacherous clay soils, before coming up alongside Lady, who was now perhaps 10 metres west of where we stopped. Lady was a rock star of the lion world. There were few more famous lions on the planet, and I was as excited as a child in Atlantis.

But excitement quickly turned to something else: Lady looked to be in trouble. Her ribs pressed through her skin, as if she hadn't eaten in more than a week. Her face was scarred, her nose bare and black, battled-scarred from years of hyena fights and dangerous hunts. Her right ear drooped flat against her head, her jowls sagged to the same side as saliva hung from her mouth and her tongue poked between her teeth. She breathed heavily.

Lady had already lived more years than most lionesses in the wild. The scientists who knew her estimated her age at around fifteen or sixteen. Although lions can live to 30 years old in captivity, wild lions rarely live beyond twelve, and almost never more than sixteen; given Lady's story, to have lived even that long was remarkable. And yet Induna Mundandwe had recalled knowing of Lady when he became *induna* in 2000, knowing that she was one of just two lions left, and that she was a fully grown adult at the time; by his calculation, she had to be at least eighteen years old. Then there was Tembo, who was adamant that Lady had been alive and well when he first arrived in Liuwa, back in 1991; that would have made her 25 years old. To the local people she was many years older than that.

Whatever her age, Lady looked close to the end of her life. She limped when she walked—a serious limp that would have made running, let alone hunting, an impossible task. 'She doesn't look good,' said Induna Mundandwe. 'She is injured. Probably from a zebra.' A flying zebra's hoof can break a lion's leg as if it were a matchstick. Out here, with Sepo secluded at Matamanene with her cubs and the males bonding in the *boma*, she was once again the last lion of Liuwa, even if only for a few weeks; she would have been hunting alone. Such was her state that I wondered whether each heaving breath she took might be her last.

Tembo and I discussed her condition, and for a time Induna Mundandwe joined us. For perhaps half an hour we spoke and watched. At one point, I turned to ask the chief a question, but he was absorbed in a weeks-old newspaper from Botswana that he had found on the floor of my car, drinking in the news from an outside world that he missed.

Close to sunset, and with great effort, Lady rose to her feet and walked with faltering steps towards the south.

'Call to her, Tembo,' I asked.

And so he did, like a lover's lament: '*Lady, lady, lady, lady . . . come Lady, come . . . come, come Lady . . . come ooooon . . . come, come, come Lady . . . don't get clever, come, come, come . . . come . . . Lady, come on!*'

I want to tell you that she came. I want to be able to say that Lady recognised her old friend and came towards us for one final meeting. But she didn't. She simply stopped and turned her head, blinked once, then blinked again in that inscrutable way of cats. Then she limped off into the tall grass.

~

There was one final missing piece to the puzzle of Lady. If Lady was in fact a person who had become a lion, the last of the Siyenge elders were the only ones who knew the real story.

Not knowing if Lady had survived the night, I sat down the next morning in the dust of Kandiana village, the home of the Siyenge. These are Liuwa's lion people, Liuwa's true lion custodians stretching back to a time when Liuwa's plains and villages, with their lion and human subjects, were a playground for Barotse kings and queens. While everything around them changed down through the centuries—as lion kingdoms rose and fell, as Barotseland fell under the sway of colonial armies and administrators, then post-colonial Zambia's technocrats and autocrats—Siyenge elders connected Liuwa to its past, to a world where the boundary between wild animals and people remained blurred by magic and superstition. Theirs was a world in which people become lions.

Fewer than 30 people lived in Kandiana. They slept upon floors of brown sand, in square huts built with a wooden frame, faced with mud that had to be reapplied after heavy rains, and crowned by roofs of wired straw. From early morning, the children of the village went to school under a tree; the schoolteacher was the only young man in the village on the day that we visited. Women worked the fields, digging furrows, harvesting cassava, then pounding millet and cooking through all the hours that the day brought. When the wind blew, as it did often, sand clouded the sky and stung the skin, and the sound was that of an approaching storm.

We sat in the sand beneath a venerable old mango tree, and our presence drew all of those in the village with no better place to be. They sat behind Dexter, who, as one of the last keepers of Siyenge secrets, was the principal storyteller. Their greeting was warm, and they were outwardly eager to answer my questions. But when they grumbled about African Parks to me, only to announce themselves to be entirely happy hours later when Rob Reid arrived with Induna Mundandwe, I was reminded how difficult it would be to unearth the secrets at the heart of Lady Liuwa's story.

When I listen now to the recording of my conversation with the Siyenge elders of Kandiana, as I have done many times in the months and years since, I am struck by the air of utter incomprehensibility that lay between us. There was goodwill on both sides, and Tembo, interpreting, assured them many times that my intentions were honourable. Yet they seemed to wonder why on earth I was asking about their ancient traditions if not to use what I learned against them in some way. And they were as incomprehensible to me as I was to them— I could barely imagine their lives, I grew frustrated with their circular way of answering, I implored Tembo to tell me where I was going wrong. In my years of interviewing, this was one of my most difficult.

At the same time there was, I came to understand only later, a strange wonder at large in that conversation, sitting on the sand beneath an old African tree, listening to elders tell stories from a past that only they understood; how much more so that they were often only half-truths told to keep safe their culture from prying outsiders. As I listened to the recording and studied the transcripts, listening as much to the silences where questions were avoided as to what was said, I came to appreciate something altogether different from what I had expected. Cast adrift from what French philosopher Michel Foucault described as 'all the familiar landmarks' of thought—'the thought that bears the stamp of our age and our geography'—I felt myself to be swimming in 'the exotic charm of another system of thought', one utterly foreign to my own. Over the hours that we spoke, as we circled and danced, we came to a certain understanding, a meeting point; it was surely incomplete yet somehow satisfying as each element emerged, as if from a chrysalis, imperfect yet enough. Kandiana was another world to me, impenetrable and dignified, utterly unwilling to sell its soul.

This is the story that they told me. It all began with the first Barotse ancestor, a man named Mbo. Mbo had a son named Susiku, and a daughter whose name was Mambeti. (There is no way for me to replicate the music with which Tembo says this name—*Mam-beti*—a double emphasis that you don't find in English.) The family came from south of Liuwa, from a village called Katongo, close to Mongu. It was a time of tribal wars, of bitter, internecine battles to rival those of medieval European armies, of escapes aided by the supernatural. When someone comes to write a comprehensive history of the Siyenge, they will find some extraordinary material to work with.

But for our purposes, the story begins at the moment when Mbo sent his children, Susiku and Mambeti, towards the north, where they became the first of their people to enter Liuwa. There they encountered lions in great numbers. Harassed by these lions, Susiku travelled far away, to Imiliangu near Sioma, a place now close to the Zambia–Namibia–Angola border triangle. It was a land where the local people had learned to live at peace with the lions around them; there Susiku searched for medicinal charms that could combat the hostility of Liuwa's lions. His journey was successful, but Susiku's son, Monde, was killed. Susiku returned to Liuwa with medicine that enabled him to communicate with lions, to send them out to kill the livestock of an enemy or to hunt wildebeest on Susiku's behalf. Liking the land, he stuck his walking stick into the ground; it struck roots and grew to become the palm tree that still stands south-west of Matamanene.

Mbo was pleased with his son's work and granted Liuwa to Susiku and his descendants. When Susiku died, lions came from all over Liuwa to the village and there they roared continuously until Susiku was laid to rest. From then until now, the Siyenge have used the charms gathered by Susiku to turn a charging lion—'Many times we have done this,' Dexter told me. And to this day, Susiku's descendants, the Siyenge, remain the custodians of the land and special companions to the lions of Liuwa.

I knew that there was more but it was also clear that Dexter and the Siyenge elders had taken me as far into the story as they could go. According to Tembo, a fear of 'witchcraft' prevented them from speaking further. I didn't notice it at the time, but when I listened later to the recording, I realised that every time Dexter was about to reveal something new and important, a gust of wind swept through the village, drowning out all conversation. I have no explanation for it other than bad luck. But like everyone touched by the story of Liuwa, I cannot discount the presence of some strange magic.

In the end, it was left to Tembo—the outsider, Mr Liuwa—to finish the tale. This he did with their blessing, on our last night in Liuwa, around a campfire as hyenas circled.

The medicine that Susiku had gathered was not just to protect the Siyenge from lions, he said. It enabled them to transform themselves *into* lions. And the Siyenge didn't send the lions to kill a cow or wildebeest. They *became* the lions, moving easily between the human and leonine worlds whenever circumstances required them to do so. But the medicine that once enabled the Siyenge to do this has now been lost, and all that remains are the charms that protect them from lion attacks. No Siyenge shall ever again turn themselves into a lion.

Tembo let this sink in. When satisfied that I had understood, he took me back to the moment before Susiku turned himself into a lion for the very first time. While still in human form, Susiku told his son, Monde, to go to a nearby pond and that, while there, he should not be afraid of anything that he should see. If he disobeyed or became frightened, Susiku assured his son, he would die. They went together to the pond and there Susiku transformed himself into a lion. Monde became very afraid and called for help. True to his promise, Susiku, in the form of a lion, killed his son. The blood of his firstborn son thereafter became the necessary ingredient that would enable the Siyenge to cross the forbidden boundary between lions and human beings.

Upon Susiku's return to Liuwa, Mambeti drank the new medicine and she, too, turned herself into a lion. She became Lady Liuwa. From then on, the Siyenge had no choice but to kill their firstborn sons if they were to turn themselves into lions. From Dexter's generation onwards, they have refused to do so. And when Lady dies, the tradition, as old as the Siyenge presence in Liuwa, will end forever.

We both sat in silence, save for the whirr of insects and the crackle of the fire. Lightning lit the sky beyond the horizon.

When I had come to Liuwa, I had imagined this would be a story of sacred lions, of clashes between old traditions and the human and conservation values of the new. I had believed that the future of Liuwa's lions would only be secured if the traditions of the past could be revived. But I now realised, not for the first time in Africa, that I was wrong. I had assumed that ancient traditions are always somehow noble, and that the holders of those traditions were diminished when these traditions were lost.

But the Siyenge knew better. They understood that, even while most traditional ways were worth preserving, some traditions should no longer survive, and that there was far greater dignity in discarding a tradition that had no place in their modern world. In the end, the Siyenge loved their children more than they loved their traditions. And in the process, the Siyenge—together with Tembo, Induna Mundandwe, African Parks and one truly remarkable lion—saved Liuwa from an eternity without lions.

~

After I left Liuwa, I thought often of Lady, and of Tembo, and of the many strange things that I had seen and heard.

Despite the terrible condition in which we had found her in early October 2016, Lady survived and, in the weeks that followed my visit, she even made the long journey back down to Matamanene where she was often seen at rest near the *boma*, sleeping close to the perimeter fence where the two males were learning to live together. A month after my visit, in mid-November, the African Parks team confirmed that Sepo had given birth to two male cubs. In a December update, the lions were doing well, and now there were nine of them. By all accounts, Lady proved to be an ever-doting aunt, helping to care for the newborn cubs, leading some to speculate that the pride she had craved during so many lonely years was finally a reality.

But as always with lions, it soon became more complicated. The two males—the son of Sepo and the new male from Kafue—never bonded sufficiently to form a true coalition. Sepo's son was most often seen in the company of Sepo, Lady and the cubs, while the Kafue male turned solitary, roaming Liuwa on his own. He was last seen with Sepo's son in March 2017.

In June, Daan Smit from the Zambian Carnivore Programme wrote to tell me that, nine months after I had last seen her, Lady was 'going strong though very stiff. I am not sure if she is able to keep up with the rest of the pride going North. But Lady has surprised us many times, so we never know. We might see her here in Matamanene in December again'. On 15 July 2017, Rob Reid saw her with her adopted nephews, digging an aardvark out from its burrow. Three days later, Charlotte Pollard from African Parks wrote to tell me that during the most recent sighting, 'Lady was limping, and we did think that maybe her time was coming to an end. But she has bounced back and, as I said, looking strong again! She just keeps on going!' Four days later, members of the African Parks team sighted her close to Matamanene.

And then—nothing. By 3 August, with no sighting of Lady in more than a week, everyone was starting to worry. The pride left Matamanene, heading north with the wildebeest herds towards the Katoyana campsite where I had stayed. Lady was nowhere to be seen. The team's attempts to find her were hampered by the limitations of her VHF collar, with its range of just 6 kilometres; because of Lady's age, the team had decided against fitting her with a much heavier satellite collar. Problems with the park's plane meant that it was a few days before the team could take to the air in a bid to find her.

On 9 August, the day before World Lion Day, Rob Reid, flying overhead, picked up a signal from Lady's collar around 15 kilometres north of Matamanene, but he couldn't see her. Marking the spot, he later returned with a ground team to the site. There they found Lady's collar, broken into several pieces. Nearby was the lower jaw of a lioness. From the distinctive groove in one of the teeth, the team confirmed what everyone had feared: Lady Liuwa was dead.

A few days later the team made another grim discovery: Sepo, too, had died. As far as the team could work out, Sepo, who had been in Liuwa since 2011, was killed by the Kafue male. Perhaps emboldened by Lady's death and knowing that Sepo and Lady had formed a formidable team but were more vulnerable alone, the male killed Sepo when she tried to defend her nearly one-year-old cubs from an infanticidal attack. Miraculously, the cubs survived, but the pride was in disarray.

Everyone at Liuwa struggled to come to terms with Lady's death. I thought of Tembo, in mourning for his friend. Rob Reid wrote an obituary in which he spoke of Lady as a lion who 'didn't look at you like a lion—there was none of that bone chilling stare, head held high, alert "through you" sort of look. She would give you . . . [a] look that had seen it all, and been through thousands of nights of loneliness'. And he acknowledged that he had heard too many stories and seen too many things in Liuwa that didn't make sense: 'Something special, something unusual, something that cuts straight across the pure science of fact.'

Data downloaded from the lions' collars enabled the team to piece together the pride's last movements before it all fell apart. On 26 July the pride—Lady, Sepo, the son of Sepo and her other cubs from two litters—killed a wildebeest around 15 kilometres north of Matamanene, close to where I had last seen Lady. Despite her limited mobility, Lady had followed the pride there, into the heart of the Siyenge's traditional lands. It was a homecoming of sorts. There she ate one last meal with her lion family. Then, her work done, Lady Liuwa—or Mambeti—lay down to rest for the last time.

There is a scientific explanation for the fact that so few of Lady's remains were ever found: out on the African plains, nothing goes to waste, and Lady's carcass was just one more meal for the hyenas.

The local people knew better. Lady, the lion that lived, was not dead. She had finally returned to her people.

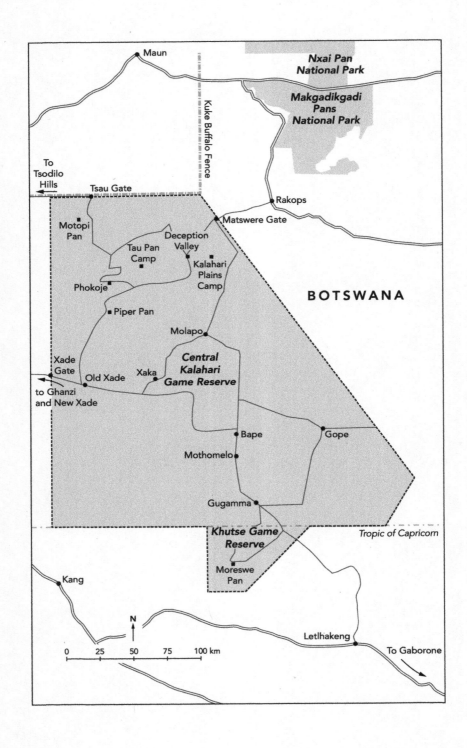

4

Kalahari Silences

Botswana, 2016

The lion roars not long after midnight, far away, in places unimaginable. At first it hesitates, the leonine equivalent of clearing one's throat; then, finding its range and warming to its theme, the lion roars again, a sound that rumbles out across the dark grasslands of the Kalahari like the drum roll of an approaching war. One part melancholy and three parts gravitas, the roar seems to rise from deep within the earth, its waves of shock and awe echoing through the night. The lion's roar has the ceremonial timbre of royalty. It fills the darkness, crystallising in an instant the loneliness of the Kalahari night. The king is on the march.

I am not really asleep, at least not deeply—the night sounds of Africa always stir in me a watchfulness that works against meaningful sleep. Perhaps this is how a lion's prey sleeps— never truly at rest, ready to run at a moment's notice.

But there is no reason for me to be afraid: the lion is far from here. A lion's roar peaks at 114 decibels, the equivalent of a rather noisy rock concert. How far that roar carries, how far away it can be heard, depends on what you are. Lions can hear other lions roar up to 8 kilometres away. Human beings can hear them across about half that distance, perhaps slightly more in the still, moist air of a cold Kalahari morning. I rather like the idea that the lion is ten times closer to me than I am to the nearest member of my own species. I can hear the roar clearly. It is close enough to be thrilling, but not too close for comfort.

When the lion falls silent, I open the canvas and gaze out into the soft green-blue night. Nothing stirs on the salt pan, luminous in the moonlight, but such a peaceful scene masks terrors that I cannot see. Were I to step outside and walk, I could soon be dead.

The night crickets resume their chorus. A long period of silence follows. I am drifting off to sleep when, at 2:14 a.m., the lion roars again. He is still far away to the north. I toss and turn but I cannot sleep. With the clock showing 3:37 a.m., his roar is suddenly louder; perhaps it is the wind. By 4:24 a.m., there is no denying that he is moving in my direction. Perhaps half an hour before dawn, he roars one last time. If the lion had been 4 kilometres away when he began, he is much, much closer now.

Unsettled, I rise before the sun to pack up my camp. Weary from having slept poorly, I am worried, and not just about the lion: I am nearly out of fuel, have very little water and don't know if I will make it to safety before either runs out. In the pre-dawn gloom, I see how precarious my position is: a high wall of grass surrounds me and if something should happen, no one will hear my cries for help. I am on my own and the lion could be anywhere.

In the morning chill, nervous as hell, I run through a list of tasks in my head, tasks that long ago became automatic routines but now seem like sacred rites of morning. My movements dulled and my hands trembling, I fumble with the latches on the vehicle and cast anxious glances all around me. Every sound and every movement on the periphery of my vision bring the morning into sharp and sudden focus. And I'll be honest: it is terribly, terribly exciting.

And then it comes.

As I teeter high on the vehicle, silhouetted against the sky while trying to batten down the roof, I hear a quick intake of breath before the lion's deafening roar rumbles through my bones. I freeze and wait for the lion to end my life.

~

On another cold, clear morning, two weeks before, in May 2016, I had stood on the hard-baked earth and looked south, a feeling of vertigo never far away. An old desert, the Kalahari—the largest expanse of sand on the planet, larger than Algeria, Africa's largest country, or Greenland—began at my feet from where it unfurled, unimpeded until it grabbed at South Africa's coastal hinterland, more than a thousand miles away to the south. In prospect, the Kalahari seemed vast and frightening, a void in the heart of a continent, and I imagined myself poised upon the brink. Of what? Perhaps it was oblivion? Or was it folly? That I planned to cross this desert alone seemed like an invitation to both.

Had I known that morning what lay ahead on my Kalahari crossing, I would surely have returned home. Or so I tell myself now. In truth, I have spent a lifetime launching myself into the unknown; as a child I jumped into the deep end long before I could swim. As an adult I have continued in the same vein, not so much safe in the knowledge that all would be well as willing to suspend both fear and careful consideration in the hope that neither would be necessary. I am the character in the cartoon who runs off the cliff and keeps going out over the chasm for as long as he refuses to look down. My assumption? That bravery will triumph over reason, and that by stubbornly moving forward I will bend fate to my own purposes. It has served me well, this philosophy of reckless optimism, carrying me—safely, it must be said—through a life freed from any shadow of regret, from having to confront questions of what might have been. Even if I had known what lay ahead, this time would have been no different.

On that May morning, not long before I stood and looked south, not far from there, I had watched a lioness march into the north. Rolling and muscular, she moved with menace, and her paws, metronomic in their rhythm, struck the dust as if calling all the creatures of her realm to attention. She certainly had mine. I have held the paw of a sleeping lion and, more than the scythe-like teeth, it is the paw that casts a lion's spell of awe; weighty and grand, it is designed to swat down rivals and prey with astonishing force.

A cheetah flows through a landscape, gliding across the grass, the epitome of wild, feline grace, lithe and nimble and quick. A cheetah is less *of* the earth than it is the earth's most fluid trespasser, and in its walk is an echo of the same fear that stalks its prey. The leopard, less alert than intent, is stealth and silence given flesh and movement, shunning the spotlight as it melts into its surroundings and disappears into the world through which it passes. The leopard prefers itself unseen, becoming one with the shadows and lying in wait until all other creatures have forgotten it is there. The walk of a lion is a statement of divine right, an open challenge to the world.

One great paw after another came down, and I watched as one might watch an epic storm or a million wildebeest on the move. When the lioness paused to stare to the horizon, she became somehow eternal, until she resumed her dawn march. I lost sight of her and now, ready to begin my Kalahari journey, I wondered if she was near.

The dry season had not long begun, and the last tinges of green were already fading. Jackals, always furtive, came and went from a nearby waterhole where vultures waited in the treetops for some calamity from which they could profit. In time, elephants came, marching in a line. When one stopped, the elephant that followed wrapped her trunk around her companion's leg; when this didn't work, she nudged her onwards with her head. Nearby, a young elephant raised his trunk in a question mark, sniffing the air, swaying his front right foot in agitation. The youngest ones, thirsty, ran the last few steps into the water, there to spray water at the jackals.

I waited, watching. I knew that once I began the journey, its momentum would carry me forward and there would be no turning back.

Why was I there? Because the Kalahari is home to an important, imperilled population of legendary lions known for their black manes and ferocious personalities; their existence is a reminder that lions, if left unmolested, can thrive anywhere, even in one of the most barren environments on the planet. Because the San people, the Bushmen, the Kalahari's First Peoples, have been expelled from the Central Kalahari Game Reserve (CKGR) and the Kalahari is falling silent. And because the relationship between lions and the San may well have been unique, perhaps even the oldest on the planet—as John Vaillant has written, the San and the Kalahari's lions 'had known one another, effectively, "forever"', and their relationship was 'calibrated long before the Pyramids were even imagined'. When that relationship ended, the Kalahari became a symbol of humankind's broken relationship with lions. I was there because I wanted to see what happens when the relationship between fragile ecosystems, people and lions falls apart, and I wanted to catch a glimpse of the last survivors. I was there to bear witness to these things, and to mourn their passing. Only deep in the desert, with its spare winds and far horizons, its silences and miracles of life in otherwise inhospitable places, could I write about those stories of death and extinction, exile and loss. Would it change anything? No, but the stories still had to be told, if for no other reason than to mark the passing of a time on earth that has forever ended.

I climbed into the driver's seat and drove south.

~

The San are an old people, as old as human time.

Some 70,000 years ago, give or take a few thousand years, a small party of San entered a cave high in the Tsodilo Hills in what is now north-western Botswana. Inside the cave, they left spearheads as offerings and, on a rock using stone tools that they then cast aside, they made a series of ritual indentations 6 metres tall and 2 metres across. This rock bears an uncanny resemblance to a python's head; in the San story of creation, humankind descends from a python, whose slithering path through the Kalahari in search of water created the valleys between its hills. The cave was so secluded that it was known only to the San until the 1990s. In 2006, archaeologists unearthed the stone tools and spearheads, evidence of what was then the earth's oldest-known human ritual, pushing back the historical record by 30,000 years, and shifting it from Europe to Africa. It was, at the time, also the earliest archaeological record of the San's presence upon the earth.

This 70,000-year-old record captures a moment in time. It does not tell us when the San arrived in the Kalahari. In the oral history of the San, there is no legend of arrival, no collective memory of making the Kalahari home. The San believe that they were there when the earth was created, and their memory-keepers speak of First Times, a period when human beings and animals were one, intermarrying and living in harmony—it sounds a lot like an oral history of evolution. 'We came from nowhere else,' writes Kuela Kiema, a San elder statesman, in his landmark 2010 memoir *Tears for My Land*. Ask any San and they will tell you the same: 'We have always been here.' Perhaps they have.

Alongside a shallow pool in the heart of the Tsodilo Hills, not far from the python cave, stands the San's Tree of True Knowledge, the Tree Where the Earth Was Born. This is the San's ground zero. They believe that this ancient Kalahari tree is where the earth began, where the creation spirit, N!adima, brought the earth into being and then knelt beside the pool. The San call one of the hills Male Hill; another is Female Hill—the San Adam and Eve, cast not from paradise and out into the world but into stone. There is even a snake at the heart of the story.

As befits the site of the earth's first creation, legends swirl around these hills—the San call them the 'Mountains of the Gods' and the 'Rock that Whispers'. The Tsodilo Hills have always resisted those who would seek to unlock their secrets. Most famously, Afrikaner writer and filmmaker Sir Laurens van der Post describes in his 1958 classic *The Lost World of the Kalahari* how, while trying to film in the hills for a documentary, all of his cameras ceased to work, his tape recorders jammed, and bees swarmed his party for three days without rest. His San guides became restless—clearly the gods were unhappy and van der Post's party was not welcome. Upon questioning the other members of the expedition, van der Post learned that, in contravention of their San guide's strict instructions to do no harm, two members of the expedition had killed a warthog and a steenbok as they approached the hills. Before he left, van der Post wrote an apology to the spirits and buried it in the hills.

I have my own story of the spirits of Tsodilo. In 2012, on my first visit, I planned to spend a few days exploring the hills. First,

I wanted to find a famous ochre painting of a lion, said to adorn a rock face on the lower walls of Male Hill. The image is one of more than 4500 paintings and engravings scattered across the hills, among them images called 'The Dancing Penises' and others that accurately portray whales and penguins more than a thousand kilometres from the sea. My approach was difficult; the track narrowed but I blundered on until the vehicle could go no further. In wild country, fearful as ever of lion and leopard, with the ringing chorus of insect buzz beating around my ears, I left the car and continued on foot. There was a sense of foreboding, a darkness of spirit that I cannot explain, but I was hot and bothered and tried to ignore my growing unease. Midway to the rock wall, a dead mongoose hung from the fork of a low tree. It was an arresting sight. A leopard? No leopard would stash its kill so close to the ground where scavengers could reach it. Nor would a leopard climb such a flimsy tree.

It felt like a warning. I shuddered.

I'd like to say that, out of respect, I returned to my vehicle and drove away. Instead I pushed on, laughing at my superstition, assuring myself that any San spirits would know that I came in peace.

I found the lion. Finely rendered in outline, the painting was one of the most accurate rock-art representations of a male lion I have seen in Africa. Its tail barely visible thanks to water damage, and in pale contrast to the vivid outline of the torso, this lion was on the run—from pursuers or in pursuit of prey, although neither was visible. Clearly the artist had an affinity for lions and had observed them at close quarters—the muscular frame, the flanks primed and ready to pounce. The image was everything I had hoped for. But having ignored the earlier warning and continued on regardless, I felt sure that I shouldn't have been there.

Restless and shaken by the whole experience, I drove to another part of the hills. As I went, my vehicle developed an inexplicable quirk of steering that I was unable to resolve, and the gravel road that led into the Tsodilo Hills disappeared from my GPS, never to return. I camped close to Female Hill. Before sunset, the wind, the *g//ausi*, which the San believe to be a lament for the spirits of the dead, whistled through the hills, and the grey go-away-bird called its eternal, descending command—*go-awaeeeey*. I felt with sudden conviction a presence of spirit, neither threatening nor comforting, perhaps human, perhaps not. Throughout the night of the full moon that followed, the presence remained with me, at first in a deep earth silence; I whispered an apology. Not long after midnight, a frightened dog began to bark and continued until dawn. I barely slept and drove away at first light.

～

Science may support the San belief that they were here when the human story began. A 2011 study found that the San, along with Tanzania's Sandawe and Hadza people, ranked as the most genetically diverse of any living people ever studied. In scientific circles, this diversity points to a close relationship between the San and the first modern human beings ever to walk the earth. Mitochondrial DNA studies have found that the San carry in their blood some of the oldest human Y-chromosome haplogroups—specifically, the 'A' and 'B' subgroupings of the haplogroups. These are the two oldest saplings of the Y-chromosome human tree. According to a 2016 set of fully sequenced genomes, the San first became the San around 200,000 years ago. James Suzman, a respected anthropologist who has lived among the San, reviewed the scientific evidence from archaeologists and geneticists, and concluded that the northern Kalahari 'may have been the birthplace of modern Homo sapiens'. He also found that the direct ancestors of the San 'probably lived in this broader region in an unbroken line from a time many tens of thousands of years before the first anatomically modern humans set foot in Europe, Asia, Australia, or the Americas'.

A 2019 study published in *Nature* strengthened this conclusion. 'It has been clear for some time that anatomically modern humans appeared in Africa roughly 200,000 years ago,' Professor Vanessa Hayes, one of the authors of the study, told the BBC. The study went on to identify Botswana's Makgadikgadi region—once a vast inland sea, now the world's largest network of salt pans, in the Kalahari's north—as the homeland of all humans alive today. The BBC's report about the study described the northern Kalahari as 'our ancestral heartland 200,000 years ago'. It was from there, the study suggested, that modern human beings dispersed throughout the world. If that is true, the San were indeed on the earth at the very beginning of human life as we know it.

The genetic diversity of the San—those who remained while other human beings migrated across the world—survived because they lived in the heart of a desert. Desert conditions diverted the great Bantu migrations that reached southern Africa around 1500 years ago and irrevocably altered Africa's demographics. The desert also kept at bay the wave of settlers that began their expansion through southern Africa after the arrival of Portuguese and Dutch ships from the fifteenth century onwards; it was a time when other peoples were wiped out by war, disease and the appropriation of their lands. Thus protected and with the stories of their uninterrupted presence in the Kalahari written in their genes, the San still carried in their blood—and to this day carry in their blood—a genetic diversity that has survived since the dawn of humankind.

~

Unlike the Tsodilo Hills, the central Kalahari, a few hundred kilometres to the south-east, preserves no time markers: sand, blown by the wind, is not an ideal canvas for dating the human presence. But we can speak of more recent times with some certainty.

The San lived in the Kalahari relatively undisturbed until the arrival of Europeans and their colonial ambitions. In the 1840s, David Livingstone passed through the region, eager to bring the Gospel and to save Africa's so-called 'primitive' peoples from the perils of lions and other wild animals; in trying to shoot a lion, Livingstone himself was seriously injured.

Two decades after Livingstone, gold prospectors began to range across southern Africa. In 1874, the flamboyant Hendrik van Zyl arrived. He was fabulously wealthy and the antithesis of San restraint—he and his sons killed more than 100 elephants in a single day, and oversaw the execution of 33 San men, women and children in retaliation for the death of an Afrikaner child. Van Zyl went on to establish the Kalahari's first cattle ranches and he is rumoured to have left a lavish treasure in Gcwihaba Cave, beyond the Tsodilo Hills to the south-west. Although few San were aware of it, in 1885 they became subjects of the British Crown, whose representatives proclaimed the protectorate of Bechuanaland, the precursor to Botswana. Settlers from the Cape Colony to the south— among them Cecil Rhodes, a colonial businessman, politician and towering figure in British South Africa—came to the region in the 1890s, especially to the western Kalahari close to what is now Ghanzi. Tswana farmers and growing settlements surrounded the San to the east.

Such encounters rarely ended well, at least for the San, who had never before gone to war. Their only weapons were better suited to hunting antelope than defending a people; they were terribly ill-equipped for the coming of the modern world. The San's world was changing, but their behaviour remained oblivious—to the point of heartwarming naivety—of the Machiavellian calculations that drove the colonial march. In the 1890s, with Afrikaner farms steadily encroaching on San land across the Ghanzi Plateau, the following occurred:

When drought overtook the advance party that had come to ascertain the suitability of the land offered to them by [Cecil] Rhodes, the men had to leave their wives and children in the cave of local Bushmen while they made a dash across the Kalahari to get help. The drought worsened and it was nearly two years before they could return. When, at last, they managed to get back to Ghanzi, they found that, despite the severity of the drought, the Bushmen had taken excellent care of the women and children.[1]

To this day, some descendants of these settlers maintain a vow to grant any San request for food or shelter.

Despite all the changes, many San simply retreated into the desert and continued to live as they always had, in a dispersed and egalitarian society that kept largely to itself. Small family bands of hunter-gatherers, many led by women, moved when the seasons changed, travelling often in search of greener pastures. Women gathered tubers, berries, *tsamma* melons, bush onions, ostrich eggs; the shells of the latter were used for water containers and jewellery. In the dry season they ate insects—grasshoppers, beetles, caterpillars, moths, butterflies, even termites—for much-needed protein. And their possessions were few: blankets, digging sticks, firewood, leather pouches. Meat was a luxury, but the San loved to hunt, killing their prey with arrows or spears dipped in a slow poison that came from beetle larvae. A successful hunter divided the meat among everyone in the band. There were seasons of plenty when everyone ate well and life was good. And there were bad years, sometimes many in a row. But it was, on balance, enough to sustain the San longer in their traditional lands than any other people on the planet.

In the San view of the world, the land was not theirs but belonged to N!adima, the creator of all things: to take more than they needed was to risk offending this capricious god and incurring his wrath, and in any case would have been unwise for a people living in such a precarious environment. The San carried with them all that they owned, they took from the land no more than they needed, and they remained largely unaffected by the world beyond the Kalahari.

It is easy to idealise the San, living as they did in harmony with the natural world and bothering no one. Of course, like any people, they quarrelled among themselves, were suspicious of outsiders and, for a time, regarded education as a foreign evil. But their shortcomings came from the past and, having yet to acquire the destructive vices of modern existence, the San were ripe for romanticising.

Laurens van der Post led the charge with a series of novels and documentaries, and his book *The Lost World of the Kalahari*. The San were, van der Post wrote, a 'vanished tribe', 'children of nature' and 'mystical ecologists'. In July 1969, a special feature in *Time* magazine described the San as an 'elysian community' where food was 'abundant and easily gathered' and the people 'comfortable, peaceable, happy and secure'. In 1980, the San became the unlikely stars of the excruciating South African cult film *The Gods Must Be Crazy*, in which a Kalahari San man finds a Coca-Cola bottle that becomes a crude symbol of San innocence. In case the point was missed, the film's narrator describes how the San 'live in complete isolation, unaware that there are other people in the world'. Echoing van der Post and *Time*, the narrator goes further, describing the San as 'the most contented people in the world. They have no crime, no punishment, no violence, no laws, no police, judges, rulers, or bosses. They believe that the gods put only good and useful things on the earth for them. In this world of theirs, nothing is bad or evil'. The San, circa 13,000 BC, make an appearance in James Michener's historical novel *The Covenant* (1980), and two San characters feature in *The Burning Shore* (1985) by Wilbur Smith. Almost universally, the San were portrayed as an ageless people frozen in time, innocent proxies for our own longing to return to a simpler, idyllic past.

Sometimes our ignorance was simply absurd. There is a predominantly San town in north-eastern Namibia called Tsumkwe. The name is an Anglicised version of the San word Tjum!ui. Had colonial officials checked, they would have learned that the new town would forever be marked on maps with the San word for a woman's pubic hair. And the San sometimes had their fun, assuring the English traveller Sir Francis Galton in 1851 that there were indeed unicorns in the Kalahari.

But our ignorance had a darker side. Since the earliest days of white settlement in southern Africa, racism and prejudice against the San were less fringe views than official colonial policy. In the seventeenth century, according to Suzman, one Reverend Father Marcel le Blanc described the San as 'the foulest and ugliest people of all the inhabited world . . . repulsive to look at and to smell'.[2] In 1799, another clergyman, a Reverend Kicherer from the London Missionary Society, swore that the San 'will kill their children without remorse' and were known to throw 'their tender offspring to the hungry lion, which stands roaring before their cavern, refusing to depart until some peace offering be made to him'.[3] Even into the twentieth century such views survived, as expressed in 1941 by Colonel Deneys Reitz, South Africa's Minister of Native Affairs: 'It would be a biological crime if we allowed such a peculiar race to die out, because it is a race that looks more like a baboon than a baboon itself does . . . We look upon them as part of the fauna of the country.'[4]

There was, it seems, no in-between. The San were either depicted as noble savages or simply savages, a shining beacon of enlightened humankind or barely human at all. Regardless of how the world imagined them, that they were able to live as they always had for so long was a miracle. It wouldn't, couldn't, last.

~

Dabe Sebitola represented one version of the San story. A guide at a luxury camp atop the pans of Makgadikgadi, in the northern Kalahari—the same Makgadikgadi where the first San had walked the earth 200,000 years before—Dabe, 35, was a new breed of San. Coaxed from the desert, he was not quite of this world no matter how hard he tried, and he was only half welcome. Aloof, world-weary, he kept himself at a distance yet hovered on the edge of conversation where his eagerness to belong, despite any outward signs of success, had atrophied into a form of disdain. There was in his profile none of the San's open, almost Asiatic cast of features, and sadness flickered in his eyes like dying embers. Even so, he resembled less a member of one of the oldest peoples on the planet than a New Age cowboy, his perfectly curled Akubra offset by a pale fleece jacket and dashing white linen scarf. At times in thrall to his own old-world authenticity, at other times showing the other camp staff that he belonged in white company, he nearly belonged to both worlds and yet claimed true membership of neither. This seemed appropriate, so assiduously did he seek to cultivate them both.

The great zebra migration of Makgadikgadi crossed the horizon, heading west; the whooping of zebras rang out across the salt pans. Before us, a small band of San led an even smaller band of tourists through grasslands that fringed the pans, gathering medicinal plants and traditional foods. At first Dabe appeared uncomfortable and kept his distance from his semi-naked kinsmen and women. As women pulled tubers from beneath the soil, men flexed their bows and made to hunt; one spry old man, wire-thin, removed his thin cloak to reveal the scars left by a lion attack many years before. As they laughed and sang in the impenetrable click language of their people, Dabe yielded, announcing, a touch too loudly, that he would like nothing more than to shed his Western clothes, dress in a leather loincloth and join them.

In one thing, Dabe was consistent: he was mortified that I was planning to cross the Kalahari alone. Although capable of perfect silence, many desert people do not like to be on their own. From the Tuareg of the Sahara to the Bedu of Arabia, from Indigenous Australians in the outback to the Sami of the High Arctic, most desert peoples whom I have met are natural storytellers, never happier than when in garrulous campfire conversations that last deep into the night. Where we in the West seek in deserts silence and solitude, people of the desert crave community and conversation: to find yourself alone would mean that something had gone terribly wrong. Which is why, of course, Dabe thought that I was crazy. He warned me against setting out and, in the days that we were together, he never missed an opportunity to try to talk me out of my journey. Yes, he was concerned that I wasn't carrying enough fuel. And yes, he worried that I might become lost. But his main concern was this: no San would willingly travel the Kalahari alone, no matter how well they knew the desert. What if something went wrong?

Dabe knew what he was talking about. He was born in the Central Kalahari Game Reserve and had lived a traditional San life, blissfully unaware of the world outside it. His family moved with the seasons, living from the land. As a child he mock-hunted, learning by degrees. As a teenager his lessons became more serious until he was ready for initiation into the realm of San hunters, leopard-like in their combination of stealth, guile and patience. He learned the watchfulness with which a San hunter pays attention to the world around him, the eternal conversation and intimacy between a San tracker and the very surface of the earth—what the San call 'the earth's face'—as they search for signs of prey.

To prove himself a hunter, Dabe killed an eland, the largest of all antelope species, an ungainly yet surprisingly fast cattle-like creature with a great hanging dewlap that swings when it runs. Having killed the eland, having delivered the mortal blow with a poison-tipped arrow, he took skin from the animal's forehead, burned it with desert herbs, cut open his own body and then rubbed in the ashes to connect his spirit with that of the eland.

In his younger years, Dabe told me, encounters with lions had been commonplace, the currency of a Kalahari childhood. As a child he heard them roaring in the night, saw their shapes moving through the scrub. He remembered how a man, crouching behind a bush while hunting, was attacked by a lion; it happened, Dabe told me, because the man 'changed his shape' from our habitually upright stance; it was not what a lion would expect. He repeated the San belief that lions ordinarily see human beings not as prey but as fellow carnivores because our eyes are in the front of our heads.

Once, while still an adolescent, Dabe was out hunting with friends when a sudden storm swept down in a fury of thunder and lightning; in San tradition, thunder is a leopard's growl and the sound it makes as it crunches the bones of its prey with its powerful jaws; lightning is the flash of a leopard's eye. He and his friends sought refuge in a flimsy shelter, a seasonal hut left in seasons past by San hunters. As they waited out the storm, not a leopard but two lions, clearly with the same idea, ran into the hut to shelter from the rain. The lions turned to face back the way they had come, blocking the doorway, not knowing that three teenagers cowered behind them.

'We don't even know how we got out. We ran! We ran!'

What did the lions do?

'We don't know, because we didn't look!'

Other encounters were more practised. The San worked in tandem with lions, chasing them from their kills, stealing some of the meat but always careful to leave some for the lions when they returned. Driving a lion from its meal is an art form: do it too soon, when the lion is still hungry, and the lion will stand its ground; wait too long and the sated lion, fat and lazy, will also resist the intrusion. Here was ancient man, scavenging kills from the great predators of the African plains.

But the bond between lions and the San goes further. Some San groups, particularly those in Namibia, consider lions to be the members of the animal kingdom most like us. Like the San, lions are social, live in family groups and inhabit home territories across which they roam, and they hunt the same prey as their human neighbours. The San have observed that, among the animals they know, only lions display human-like affection for each other. And the San believe that there is an unspoken but entirely mutual pact of non-aggression between lions and the San—just as the San sometimes leave part of their own kill for the lions, the San believe that lions do the same.

San stories tell of male hunters being seduced by lionesses into joining the pride. Dabe told me how a San spirit man or man of healing, a *xumkosi*, can enter a trance-like state when hunting and transform himself into a lion. I told Dabe that I had heard such stories. 'It's not a story,' he said. 'It's something which San people can do.' In the northern Kalahari, Dabe and his group once tracked a man whose tracks turned into those of a lion. They followed the tracks, which soon returned to human form, and there they found the man.

Dabe had one last story to tell, from a time when 'all the animals, plants, everything, used to talk'. But the story was, he said, one that he could only tell under the cover of darkness. To tell such a story during daylight was to tempt fate, to invite malign spirits to take possession of storyteller and listener alike. Some San will not use the proper word in their language for 'lion' at all, daylight or not, for fear that a lion might hear its name and, thinking itself summoned, come and eat the person who speaks. For this reason, I withhold the San word for lion here, just in case you find yourself reading this in lion country and decide to practise how it sounds.

And so we came together again after dark. This is the story he told:

> A lion and a Bushman were once very good friends. But one day while they were hunting, the lion lost the San. He lost his friend. They became separated and the world was such a big place that there was no way that they would meet again. So the San made a fire. Then he took the ash and the coals and threw them into the sky to make this beautiful lane, which we now call the Milky Way. If you look up into the sky, you will see the sign of Leo, and that is what the San is seeking. He's missing his friend, looking for the lion. At the end of Leo, that's where the San was waiting. So the lion followed the Milky Way all the way up to Leo and that's where he found his friend. Since then, they have never again been apart.

~

In 1961, with Botswana under colonial rule and independence still five years away, the British government created the Central Kalahari Game Reserve (CKGR). At around 52,000 square kilometres, it was at the time—and remains—one of the largest protected areas in Africa. It is larger than Belgium, the Netherlands or Switzerland, nearly six times larger than Yellowstone National Park in the US, and three-quarters the size of Tasmania.

George B. Silberbauer, the man responsible for creating the reserve—he was district commissioner for Ghanzi at the time—would later write that the purpose of the CKGR was 'to protect the Bushman inhabitants of the area'. In the late 1950s, he wrote, 'illegal hunting by non-Bushmen from outside the area posed a serious threat to the hunters and gatherers who depended on the game herds for part of their livelihood'.

The colonial authorities may finally have concerned them-selves with conservation and with protecting the San through the creation of the reserve, but Botswana's ruling class, those who would lead the country after independence, were far more interested in profits. With the discovery of diamonds in Botswana still a decade away, turning a profit meant establishing cattle ranches across the wide, open grasslands of the Kalahari. There was one problem: no one would buy Botswana's beef as long as Botswana's cows grazed with wild buffalo herds where foot-and-mouth disease was rampant. In the 1960s, Botswana's rulers began to build fences to separate wild animals from livestock. These fences worked for the coun-try's economy—the system, still in place today, ensures that certified-healthy beef from north of the fence is an important contributor to the country's export revenues. But the results were catastrophic for wildlife.

As recently as fifty years ago, James Suzman wrote in 2017, San communities in the Kalahari 'still remember watching herds that would stretch far beyond the horizon and take as long as three or four days to pass'. Suzman and others have written of how more than 70,000 wildebeest, springbok and hartebeest died in great numbers when a new fence blocked their seasonal migration to Lake Xau, northeast of the Central Kalahari Game Reserve. Populations of these species never recovered.

It was not just the animals that suffered. The San survived in delicate balance with wildebeest and other prey species, with lions and with the land itself. The sudden decline in wildebeest and other prey animals seriously impacted upon the San's ability to hunt. Lions, too, were affected and, with far fewer prey animals to sustain a large lion population, their numbers fell. With lions and the San fighting over ever-smaller scraps, they went from being co-dependents who sometimes worked together to competitors, each fighting for their own survival. By building fences, by messing with a balance that been in place for millennia, the government made the entire ecosystem unravel. It marked the beginning of the end for the oldest hunter-gatherer human society on the planet.

In 1997, three decades after the government began ring-fencing the Kalahari, the Botswana authorities started evicting the San from the CKGR. There was deep division within the San community about the move. On one hand, the government was offering free land and cattle as compensation in new settlements—New Xade near Ghanzi, Kaudwane in the south—that the government established just outside the reserve. Thirty years of fences had made life increasingly difficult inside the CKGR: there were far fewer animals to hunt, drought stalked the Kalahari with what seemed like increasing frequency, some San had turned their hand to commercial rather than subsistence hunting, and younger San were leaving the central Kalahari like never before.

At the same time, the San had lived there an unimaginably long time; they knew no other life and, in the view of many, had no pressing need to leave. Life was difficult, yes, but it had ever been thus in the Kalahari. The inducements offered to start new lives meant nothing to a people with no concept of land ownership and who had never raised cattle; with no little disdain, the Tswana and Botswana's other Bantu-speaking peoples called the San *Basarwa*, which means 'Those who do not rear cattle'.

There are many theories as to why Botswana's government evicted the San. Government officials claimed that it was for the protection of the CKGR's wildlife—hunting had, they said, reached unsustainable levels and the CKGR's larger animals, already much depleted, would disappear entirely if nothing was done. The thoroughly modern Botswana ruling elite also made little secret of their opinion that the San's decision to live in the Kalahari was a lifestyle choice, one for which the rest of the country was paying: the San had, for decades, relied on government-funded health posts, water supplies, fuel and transport. And besides, the government said, it was for the San's own good. Echoing the racist stereotypes of the past, in justifying his government's decision to expel the San from the CKGR Botswana's then vice-president Festus Mogae described the San, according to the *New York Times*, as 'Stone Age creature[s]' who 'must change. Otherwise, like the dodo, they will perish'.

There was one more reason. In the 1980s, prospectors discovered diamonds at Gope, close to a San settlement of the same name in eastern CKGR. The government denied any connection between the discovery and the evictions, but few doubt that it played a role. Perhaps the government was trying to head off any legal claims from the San, as erstwhile occupiers of the land, to the diamonds and other resources thought to lie beneath the Kalahari's sands. Now called Ghaghoo, the mine was sold by De Beers to Gem Diamonds in 2007 for US$34 million. In approving the mine, Botswana's government forbade Gem Diamonds from supplying water to the San. Gem Diamonds later claimed that the mine could be worth $4 billion, spent $85 million developing the site, then sold it to a Botswana firm for just $5.4 million. As of 2019 the mine lay dormant, a victim of what one Botswana's newspaper called the 'Bushman curse'.

In 2002, five years after the initial expulsions, the government evicted a further 650 San from the CKGR. In the same year the San, backed by the international NGO Survival International and led by a San elder named Roy Sesana, lodged a case in Botswana's courts, seeking a declaration that their eviction from the CKGR was illegal. More than a third of those evicted in 2002 quickly returned to the CKGR in defiance of government orders. The government returned in 2005 to finish the job, evicting all the San it could find.

Once expelled, most San struggled to adapt to their new lives and the inevitable social dislocation that ensued. Taking advantage of a people with no understanding of legally binding contracts or the monetary value of cattle, opportunists from beyond the San world stripped many San of their newly acquired assets. Alcohol dependence became rife, stalking the squalid, bleak, made-up new towns. Diseases such as diabetes, which the San had never known, became widespread. Nearly every San family has its story of traditions and generations forever lost.

'For those of us who were raised in the Kalahari,' Dabe Sebitola told me, 'that's where our spirits are, where our fore-fathers were buried. You have to be in that spiritual place to connect with your ancestors, in order to heal.'

In 2006, Botswana's courts ruled in favour of the San, declaring their eviction to be 'unlawful and unconstitutional': they could return. But it was a Pyrrhic victory. Undeterred by their legal defeat, the government filled boreholes with cement and refused to provide health services within the CKGR. The San took to the courts once again and setbacks followed, but in 2011 the highest court in the land, Botswana's Court of Appeal, ruled that the San must be allowed to access their boreholes and were permitted to sink new ones inside the CKGR. The judges described the government's treatment of the San as 'degrading' and the San's struggle as 'a harrowing story of human suffering and despair'.

Botswana's government has been a sore loser. At every turn they have sought to circumvent legal rulings and prevent the San from returning. They removed a clause in the country's constitution that specifically protected the San and their rights. They introduced a one-month permit policy for those San who wished to return; when the San took this to court to demand free and unlimited access, the San's longstanding barrister, British human rights lawyer Gordon Bennett, was barred from the country. The government has consistently refused to provide health care, schools or any other services to those San who have returned to live inside the reserve. And they have banned the San from hunting inside the CKGR without a difficult-to-obtain and highly restrictive permit. In a related High Court ruling in 2015, the court warned that banning the San from hunting in the reserve was 'tantamount to condemning [them] to death'.

At the time of my visit to the Kalahari in 2016, and despite their legal right to return, only a few San remained inside the reserve, restricted to impoverished settlements with access to water but very little else. The government may have lost the legal fight, but the San's victory proved to be as empty as the Kalahari was without them.

~

With Dabe Sebitola's warnings ringing in my ears, I scoured the truck-stop towns along the Kalahari fringe on my way south, looking for jerry cans to carry more fuel, just in case. But the day was a national holiday, everything was shut and I resigned myself to hoping that Dabe was wrong. Soon after passing the village of Rakops I left the paved road—all going well, the next asphalt would be almost two weeks away—and drove along a sandy trail towards the reserve.

If uncertain petrol supplies darkened my mood, memories of previous visits to the Kalahari hardly improved matters. Four years earlier I had approached the reserve's northern Tsau Gate along the Kuke buffalo fence, one of the fences upon which Botswana's cattle industry relies. Coming over a gentle rise, I had surprised an eland. Feeling itself trapped, and wild-eyed with fear, it had run at the fence instead of retreating south into the reserve, recoiled in shock, then launched through the wire and emerged, limping and bloodied, on the other side.

Soon after, at Tsau Gate, I had tried to strike up a conversation with a park ranger while he finalised the paperwork that would allow me to enter the CKGR. He had been friendly enough until I'd asked if there were there any San still living inside the reserve. He had paused, the hint of a scowl playing across his face. His hand holding the entry stamp had hovered over my papers. It only lasted a second, but his mood had changed. Unsmiling, he had stamped my permit and pushed it back across the desk, then stood in the doorway and watched me go.

Hours later I had set up camp in wild country, in thick bush and near-total darkness, having missed the turn-off to my campsite. There I passed a fitful night, fearful and badly shaken.

Now, four years later, west of Rakops and en route to the reserve's eastern Matswere Gate, I surprised a puff adder, one of Africa's deadliest serpents, as it inched sluggishly across the track. I swerved and missed the snake. It reared with frightening speed.

My nerves frayed, I arrived at the gate, paid my dues and entered the reserve without further incident.

In the northern reaches of the CKGR, the earth was tinder-dry, scalded by the sun. By late afternoon, yellow turned to gold as I arrived in Deception Valley, a long, low valley of grasses that bent and swayed in a gentle breeze. Gilded sand dunes lined the valley's shoreline. Impala grazed across the valley floor. One of the most beautiful corners of the Kalahari, Deception Valley cuts through the desert like a seam of gold.

To understand Deception Valley and all such fossil valleys of the Kalahari, you must first imagine the desert as one of those handheld games where you try to steer a tiny metal ball into a hole by tilting the surface; the merest shift in the angle changes everything. For millennia, waters flowed through the Kalahari, forging valleys and filling lakes. Around 15,000 years ago, the San would have felt the earth tilt as tectonic plates shifted beneath them. It was only subtle, but enough to divert the waters away from the Kalahari and towards the Zambezi River. All across the Kalahari, rivers ran dry, lakes became salt pans and water seeped into the earth. To this day, beneath the western Kalahari is the world's largest non-glacial lake, as well as sufficient water to supply all of Namibia's water needs for the next 400 years.

The San tell a slightly different story, believing that Deception and other river valleys were created when G//aua, a mischievous god, stooped to shit upon the sands of the Kalahari. Beneath where he crouched lay a none-too-pleased puff adder. Rising in anger, the puff adder sank its fangs into the god's testicles, which swelled to many times their normal size. As G//aua wailed and dashed about the land in agony, he dragged his hideously swollen testicles behind him, gouging out valleys like Deception between the sand dunes.

I felt a special sympathy for G//aua. Four years before, I had slept on a gentle rise above the valley. Near the end of a night disturbed by warthogs grunting around my vehicle, stomach cramps had sent me running out into the open to shit beneath a tree in full view of predator and prey alike; mercifully, no puff adder had chosen to spend the night beneath the tree. Only a crimson-breasted shrike, a bird sporting one of the most glorious colours in nature, eyed me quizzically, then flitted over to investigate when I ran for cover. To this day I shudder at the thought of what it found. By deciphering the footprints, I later learned that a pride of lions had rested under an adjacent tree just before my night-time dash.

Embarrassment aside, I treasured Deception Valley, and I was not alone. 'For me, Deception Valley is eternity,' Dr Helena Fitchat, a Czech-South African conservationist, once told me. 'I have spent a long time trying to find the right word, and it is eternity.' I have always taken her at her word, convinced that this has nothing to do with the fact that she once broke down not far from the valley with no water and only a bottle of whisky for sustenance. After being spotted by a search plane she was discovered lying under her car singing childhood songs in her native Czech.

More famously, one of Deception Valley's acacia islands was, for a number of years, home to Mark and Delia Owens. Controversial figures, the Owens were expelled from Botswana in the early 1980s, partly for their international campaign to denounce the building of fences around the Kalahari and the mass die-offs of wildebeest and other wildlife that resulted. Later they left Zambia under a cloud, amid accusations that their anti-poaching teams had carried out at least one extrajudicial killing. But before all of that came to pass, they pioneered research into brown hyenas and Kalahari lions, and their warm classic of Kalahari life, *Cry of the Kalahari* (1984), is, among many things, a love letter to Deception Valley. At one point in the narrative, a Kalahari old-timer tells the Owens, 'I've been this far and no further. Beyond here, no man knows.' In all of their years in the Kalahari, the Owens never drove into the deep south of the Kalahari; the mere thought of doing so gave them pause.

Now, on my return, Deception Valley was, if not paradise lost, then at least long-ago discovered, and more vehicles than antelope crowded the valley. The south, the great expanse of wilderness that once belonged to the San, was where I was going.

~

Beyond where Deception Valley disappeared into thick scrub, I left the main trail and took faint paths that tracked south and then east, meandering along shallow valleys that no other vehicle had passed that day. Close to sunset, with the light failing, I sidled down off a hillside and skirted a large pan where, in a quiet corner of the desert, waited Kalahari Plains Camp.

Run by Wilderness Safaris, Kalahari Plains has been at the epicentre of the international war of words surrounding the San's expulsion from the CKGR. The luxury camp was opened in 2009, twelve years after the San were first evicted from the reserve, and five years after Botswana's highest court found in favour of the San's rights to live on their traditional lands. According to Survival International, Wilderness Safaris should never have opened the camp without first consulting the San. Using the camp's swimming pool as a symbol in a land where the government had actively prevented the San from drilling boreholes, Survival International called for a tourist boycott of Botswana in general, and Wilderness Safaris in particular.

The initial response from Wilderness Safaris in 2010 was that it had won the right to build the camp in a public and transparent tender process, that they had, where possible, avoided occupying core areas where the San still lived, and that all other issues, including access to water, were 'a matter between the Government and the Bushmen communities'. Then, and in the years since, they have pointed to the benefits that accrue for local San communities, and the nation as a whole, from tourist initiatives of this kind.

In Kalahari Plains Camp as a guest of Wilderness Safaris, I was reminded that the San did indeed live within the boundaries of the CKGR, and not just in the remnants of old San settlements. They do so in luxury lodges, like Kalahari Plains, there to re-enact 'nature walks' or 'Bushman walks', to play their one-stringed lutes as a form of welcome to arriving guests. There is, of course, an argument to be made that, in the absence of any ability to lead a traditional life, performance-work of this kind is preferable to life in a squalid resettlement camp. By replicating some form of cultural authenticity, these role-plays, and these forms of tourism in general, help to keep traditions alive and enable the San to earn money from one of few commodifiable resources they possess and in which the world has an interest. Just as promoting genuine San handicrafts for sale to tourists can maintain artisan traditions that might otherwise be lost, San cultural performances can help to secure the importance of learning traditional rites and customs for a whole new generation of San. And such jobs were in great demand.

But there is a hint of elegy in what they project, a reminder of what has been lost. At one point, as the small band of San at Kalahari Plains rested between performances—life for them now *is* performance—a giraffe crossed the plain in the middle distance. All of the San turned as one to watch. Slowly and deliberately, the family patriarch picked up an imaginary bow and arrow, took aim, then released the make-believe bowstring and watched the make-believe arrow arc towards the giraffe.

For the briefest of moments, he was ancient man, captured like some precious daguerreotype from the past. It was not done for me: he didn't know that I was watching. It was an act of longing and of nostalgia. Inside the reserve, San hunting—which, James Suzman wrote, 'animated [San] men's relationship with the world around them, gave them purpose, and imbued the cosmos with a sensate touch of the real'—was no longer allowed except under severely circumscribed rules. And yet, for a fleeting moment, as the man mimed the act of drawing back a bowstring, he was the proud hunter of his ancestors.

As the tourists gathered, he smiled sadly and returned to his role as entertainer. The time of the San had passed, perhaps forever. This man was now a museum piece who would never again hunt a giraffe, and this realisation was one of the saddest moments of my journey. With a vast Kalahari pan of sere grasslands sweeping out to the setting sun, the camp here was a prison and the man and his family were as much captives as freaks in some outmoded circus. And I was one of their jailers.

~

The Kalahari male lion is one handsome individual, considered by many to be king among lions. Most scientists agree that he has the most beautiful mane of any lion on earth.

Studies in the Serengeti by Dr Craig Packer, using stuffed dummy lions created for the purpose, revealed that the larger, darker and thicker a lion's mane, the more likely it was to impress a female. Lions with big, dark manes produce more testosterone, they live longer and their cubs are more likely to survive into adulthood—in other words, genes that any lion mother would want for her offspring. The mane of the Kalahari male is luxuriant, a fulsome extravagance, and it may just be this outward sign of virility that has given him a reputation for fierceness—he is indeed a fine-looking animal. Alone among lions, his mane is consistently dark, trending to black—with an affecting blond ring around his face. Although the Kalahari lion is no bigger than lions elsewhere, this fine head of hair can make him appear so.

It's not just looks. The lions of the Kalahari also have smarts. Take hunting, for example. Each time Kalahari lions set out to hunt, they can do so with more confidence than any other lion known to science: they have a 38.5 per cent chance of being successful. If that doesn't sound like much, remember that this is a land with very little cover, and that the figures elsewhere—23 per cent in Tanzania's Serengeti, 15 per cent in Etosha National Park in Namibia—are significantly lower.

Then there's this: one group of lions in the Makgadikgadi region of the Kalahari almost seems able to tell time and to calculate distance. In one study, the Makgadikgadi lion only approached human buildings between 8 p.m. and 6 a.m. when humans were indoors. Then, when they did have to pass close by, they accelerated, breaking into a run or a trot when they reached a remarkably precise 6 kilometres from the nearest building. Inside national park boundaries, these same lions were relaxed around people. Outside, they made extraordinarily sophisticated decisions about how to avoid humans.

The Kalahari female lion has her own claim to fame: she is like no other lion we know. Lions live in prides, and the basic building block of the pride is a multigenerational sisterhood of blood-related lionesses. The structure can fray around the edges but the principle generally holds. Only in the Kalahari has the structure been known to fall apart entirely. During the droughts that commonly afflict the region, the Kalahari female takes her cue from the males and becomes nomadic. Nothing so unusual there. But out here in the Kalahari she hangs out, co-operates on hunts and even acts like a sister with unrelated female lions who would be sworn enemies elsewhere; in other places in Africa, females from rival prides will sometimes fight to the death. Even when conditions improve, a Kalahari lioness sometimes sticks with her newfound friends. It's the lion equivalent of shaking family values to their core.

Despite her good looks, intelligence, and social adaptability, a Kalahari lioness doesn't have it easy. From April to October, the dry season, prey is scarce, and she must take what she can; in lean times, porcupines make up a quarter of all lion meals in the Kalahari. Otherwise, she takes down springhare, mongoose and aardvark, all of which would be considered little more than a snack to lions in more plentiful habitats.

Sometimes, she doesn't eat at all. Lion researchers once followed three Kalahari lionesses for seven consecutive nights. They walked all night, every night, covering more than 80 kilometres; Kalahari lions have been known to cover 20 kilometres in a single night. Their only successful 'hunt' during that period was an ostrich egg. Another time, three young males killed only a bat-eared fox and three porcupines over a nine-day period.

When times are good, the Kalahari lion particularly likes gemsbok, a large antelope; I have tasted gemsbok in Namibia and it is indeed one of the tastier game meats. Such a meal doesn't come easily: lions must dodge the gemsbok's deadly scimitar-like horns, perhaps the most effective defence weapons in the animal kingdom. When cornered, a gemsbok backs itself into a cul-de-sac and confronts its attacker. The clever lion will grow weary and leave. A scientist in the Kalahari once came across a gemsbok walking around with a dead leopard impaled on its horns.

Food may be scarce in the Kalahari, but water is almost non-existent, and the surface temperature of the sand can reach 70 degrees Celsius. Like lions everywhere, Kalahari lions rest during the hottest part of the day, the better to regulate their body temperatures. In the Kalahari they have learned to drag their kills into the shade and, when hunting, they conserve energy by waiting alongside trails to ambush prey when it passes, rather than stalking then running after it. Nowhere in the world of Kalahari lions is there permanent surface water, except after unusual rains of the kind that may not happen at all during a lion's lifetime. And so it is that Kalahari lions can go for weeks, perhaps even longer, without drinking water, obtaining the necessary fluids from the blood of their prey, or by licking dew from the grass in cooler months when overnight temperatures plummet close to freezing.

The survival of lions in one of Africa's harshest environments and the adaptations they've made in order to persist there have contributed much to our understanding of lions and how the species might survive in an increasingly hostile world. But the lions of the Kalahari matter for another more simple reason. There are nearly 1000 lions spread across the Kalahari's reserves—the CKGR, Khutse Game Reserve, Makgadikgadi Pans National Park, Nxai Pan National Park and the Kgalagadi Transfrontier Park—and one in twenty of all lions in Africa are Kalahari lions. That number increases if we expand the definition of the Kalahari to include Namibia's Etosha and Zimbabwe's Hwange national parks.

For lion populations across Africa, the world is closing in and changing. In the CKGR perhaps 420 adult lions remain, a fraction of what was once sustained there by the vast herds of wildebeest and other prey. Those deep in the core of the reserve are probably okay. But farms and cattle ranches crowd the reserve's perimeter and those lions with territories along its outer fringes are in trouble: a 2014 study warned that 50 CKGR lions are shot by ranchers and farmers every year.

~

I tracked west from the Kalahari Plains camp, searching for Tau (Lion) Pan, but I saw no lions—I have seen lions in the CKGR in the past, but their absence on this journey was starting to worry me. I was close to the western boundary, the area where entire prides have been shot by ranchers; they have learned to be wary.

At Tau Pan Camp I spoke with Scupa; his real name was Majwagana Tshuruu but everyone, even his mum, called him Scupa. Born in Molapo inside the CKGR in 1966, Scupa had, like Dabe, spent more than half his years living a traditional San life, and his childhood memories seemed to come from another world. As a child he ran, terrified, into the bush the first time he saw a motorised vehicle. Instead of telling him ghost stories, Scupa's family threatened that, if he misbehaved, the police would one day come for him and carry him away. Or they cautioned him not to wander off into the bush, for there lurked lions who would, they said, 'stalk, pick a person, then run away'.

Real life was soon frightening enough that there was no longer any need for such stories. One night while still very young, Scupa was sleeping alongside the campfire with his parents. He woke just as a male lion leapt over him, to snatch in its jaws one of the family's goats. He pulled his gemsbok-skin blanket over his face in terror, but his parents woke to the goat's cries of distress and tried to follow the lion; it escaped. The next morning, Scupa was foraging for firewood with his mother and siblings when his brother shouted a warning: 'Scupa! A lion! In front of you!' It was barely 5 metres from where he stood. 'It was going to eat me. I ran! I climbed up the tree!' His brother went up another tree nearby, but his mother, with a baby on her back, was struggling to climb. They began to shout, the men of the camp came running with spears, and the lion retreated into the bush.

Scupa's eyes shone at the memory. But for all such stories, he was unusual among his people: he believed that the government had been right to expel the San. His people, he said, needed to be saved from themselves, as they had turned to commercial hunting—hunting for profit—rather than continuing the San tradition of killing only what they needed. Hunting in 4WD vehicles or on horseback, using high-powered weapons, a new generation of San hunters was wiping out the Kalahari's wildlife in a manner that was no longer sustainable, Scupa told me. He knew this because he had been one of those hunters, and had participated in the wholesale slaughter of giraffe and other species, including, on one occasion, 70 gemsbok in a single month. They even hunted lions.

Scupa's confession brought to mind something that Dabe Sebitola had told me: 'When rifles were introduced, that is when the relationship between lions and the San people became disconnected. From then on, lions never trusted man again.'

A headstrong young man at the time of the expulsion, Scupa had done well in the years since. He worked as a spotter for tourists in a luxury lodge, and he owned cattle back in New Xade, the settlement created in 1997 for the relocation of the San. His family, too, had benefited. A younger brother was a wildlife ranger in the CKGR. A cousin worked for the government in London, a nephew was a police officer close to one of the diamond mines at Jwaneng in the country's south-east. His children went to school; he, Scupa, was unable to read.

'I don't want to go back,' he told me simply. 'New Xade is my home.'

According to Scupa, there were now just three populated settlements inside the CKGR—Molapo, Metsimanong and Mothomelo. 'But they are not living like the olden days. They are wearing modern clothes like me.' Then Scupa's voice dropped. 'Out in the south-east of the reserve, you might see some of those who are still living in the old way. They are still dressing like before, living off the land. After Mothomelo, east of Gugamma.'

And the government has never tried to catch them?

'They didn't see them. At the time we were kicked out, they hid themselves.'

Like his family used to, when he was a little boy?

'Like we used to.'

~

Leaving behind safari trails in favour of ragged tracks that snaked through the sand and around deserted alkaline pans, I drove south. A secretary bird, with its eagle's torso and absurdly thin legs, picked its way through the grass, looking for snakes.

A kori bustard, Africa's largest flying bird, frog-marched by the roadside. And with each passing kilometre, free from other human beings and their noise, I felt my mood lifting as if light were flooding in.

At Phokoje Campsite, surrounded by pans and grasslands, tall grasses concealed all manner of mysteries—the kind that could burst from hiding at any moment to end my days. But, in time, I was overtaken by the silence of such a remote place. As I went about my business, a Kalahari scrub robin went about hers, flitting through camp in her busy quest for water and insects. At sunset, I sat on my vehicle's bonnet and, in the greatest happiness, watched grasslands turn golden as springbok, gemsbok and red hartebeest grazed in peace. To advertise his presence to potential partners, a male hartebeest climbed a gentle rise and stood, erect and magnificent. From my own perch, I understood the inclination: I was unaccountably happy and wished the world to know it.

Later I drank whisky from a tin cup, mindful of the tall grass but content. In different times, George Silberbauer wrote of how he and his travelling band of San were harassed over seventeen consecutive nights in attacks by 38 different lions. It could happen, I guessed, but tonight at least I was unafraid, and sat for long hours in silence. I heard no evidence of lions.

～

The next morning I continued south, deep into what was once the land of the San, and my spirits soared with each passing

248

kilometre; I saw not another vehicle the whole morning. At Piper Pan, one of the loveliest pans in the whole Kalahari—like a lake where the water has suddenly retreated—I saw cheetah and an extravagantly horned kudu in the woodlands that rose from the shoreline. I drove around in large circles for the sheer joy of seeing the pan from different perspectives. It was a high point of my trip.

And then things started to go awry.

My vehicle's fuel gauge digitally calculated remaining fuel reserves in the tank and predicted the number of kilometres that these reserves could carry me. I began the day in Phokoje with plenty of fuel and an expected buffer of hundreds of kilometres up my sleeve. I even allowed myself a smile of satisfaction that Dabe Sebitola had been wrong, and that I would arrive at the next petrol station beyond the reserve with nearly half a tank to spare. Soon after leaving Piper Pan, the going became tougher, through tracks of deep sand. My fuel tank had made its assumptions based on trails—asphalt or, at worst, gravel or firm sand—where the road offered little resistance. All of a sudden, for every kilometre that I drove, my fuel-reserve estimate dropped by a factor of three. If this continued, I would run out of fuel long before I had crossed the Kalahari. I was in trouble.

I had long dreamed of arriving in Xade (often called Old Xade now), one of the epicentres of San life in the Kalahari pre-1997 and an important destination on my quest. It marked the point where the southern reaches of the CKGR, the true San heartland, began. And it was here that the San had lived in greatest numbers before they were expelled in 1997. But by the time I arrived, the San were long gone, and my journey was suddenly in crisis. At Xade Gate I asked sullen park officials for fuel, but they had none to spare. I also asked for directions to the remains of the old San settlement, but they assured me that none remained. Careworn modern buildings, a fenced electrical substation, staff headquarters—beneath these lay buried the last physical evidence of the oldest hunter-gatherer society on the planet.

Dabe Sebitola had returned as a guide to Xade just two years before my visit. It was the first time that he had been back: 'The whole area was abandoned. It really felt like I was lost. I remembered the old days as I was approaching this valley. I could smell the fire in the distance, and I really missed that as I was going down that valley. It was just a big empty place.'

Camped not far from park headquarters from where I could hear the officials' radio until deep into the night, I sat with a map, calculating fuel levels and weighing my options. Beyond Xade there would be nothing and no one, and many of the sand tracks were even deeper than those that had confounded my fuel estimates earlier in the day. This was the last place where I could make a dash for fuel and return to the trail, although the eight-hour-return detour to the fuel stations of Ghanzi would work only if I was able to find a spare jerry can. To reach Ghanzi would mean passing through New Xade, the settlement set up to house those San relocated from where I now was. And it would be a defeat of sorts, like claiming to cross the Kalahari with a timeout in the middle, breaking faith and breaking a spell. At the same time, running out of fuel in the heart of the desert, days from help, was no joke. If I continued with my current reserves, such an event was more than possible—it was likely.

I had two choices, and neither was very appealing.

~

I had no desire to return to New Xade. Earlier in the year I had visited from Ghanzi in the company of Omphile Gabobonwe, a 26-year-old San guide. Before meeting him I had driven through D'kar, a predominantly San town off the Trans-Kalahari Highway, north of Ghanzi. Within sight of a San museum and arts centre, old men had lain sprawled in the gutters surrounded by empty bottles of cheap whisky, and fights broke out between younger men who swayed, dangerously drunk. Further south, in Ghanzi, an old San woman in rags in a shopping centre car park had grinned uncomprehendingly at the world while the town's children poked fun at her.

An estimated 90,000 San remain in southern Africa, spread across Botswana—home to more than half of all San—as well as Namibia, Angola, Zambia, Zimbabwe and South Africa. Although there are individual exceptions—Scupa's family, for example—more often, where the San have collided with the modern world beyond their traditional boundaries the outcome has been calamitous. One of few peoples to have survived into the twentieth century without need of outside assistance, they became, within a generation, a subject people on the margins of the modern world. One 1998 study by James Suzman on behalf of the European Commission found that the San 'were by far the worst off of any population group in the region'; that fewer than 10 per cent had any meaningful access to their traditional lands; that the San suffered 'routine racism and prejudice'; and that they were, as a people, 'caught in a death spiral of poverty and marginalisation'.

As I had prepared to visit New Xade, the owner of one of the tourist camps, a long-time friend of the San, had advised me to make my presence known to the village headman upon arrival. 'Make sure you get there early,' he had suggested. 'Otherwise he'll be too drunk to speak with you.'

If ever any metaphor spoke to Botswana's vision of the San future, it lay in the roads that connected New Xade to the rest of the world. The road in from Ghanzi was well maintained, graded regularly and passable year-round, even after rains. Beyond New Xade, the road that joined the modern settlement with the CKGR was one of the worst in Botswana.

It had been a Sunday morning when I visited, and New Xade was quiet. Its orderly grid of wide streets crouched upon the desert fringe; New Xade was the colour of grey sand. The few trees produced little shade. Even the camelthorns, perhaps the Kalahari's most widespread trees, were more stunted there than in the reserve. Goats foraged for meagre pickings in the dust; there was more litter than grass. New Xade was ennui and unresolved transience, its inhabitants waiting for a future that few believed in. Meanwhile, the first San generation to grow up here without firsthand knowledge of desert living was nearing adulthood.

With its population of close to 3000, New Xade was divided into wards, each set aside for different San communities.

These reflected old CKGR demographics—those with origins in Old Xade, Molapo, Metsimanong and Mothomelo—as well as a Tswana ward for government workers and other non-San. Limp Botswana flags hung from flagpoles in the still air, presiding over shuttered government buildings that were surrounded by high-wire perimeter fences. New Xade was a welfare town, where a government job was one of very few means of gainful employment, while the rest of the population subsisted on government handouts; no one was starving, although crime was, I was told, commonplace. In many thinly fenced family compounds, brick government-built homes faced off across the sand with traditional, circular San shelters of thatched roofs and tightly woven branches for walls.

Born in the reserve, Omphile Gabobonwe was seven years old when his family was expelled and herded onto trucks with all their worldly possessions. Like many younger San, Omphile burned with nostalgia for a life he could barely remember. 'We were free to roam around where we pleased to go. We were free to hunt. As a little boy I used to love it when my father told me that "This is how you have to grow up and fend for your family. You have to be a good hunter."' It hurt that he had never been allowed to live out that promise, that his was the generation where the rupture had occurred. More than that, he denounced the fatalism of his elders, of those who lamented that the old ways had gone forever but did nothing to get them back. In Omphile's telling, the older generation walked away without a fight, surrendering future generations to a wider world where they encounter racism and disadvantage at every turn.

'It wasn't just any other land to us. When my parents had that land, they believed there was no suffering. They had access to the medicinal plants. They had access to the wild animals for hunting. But when I go there now, I am not feeling that vibe, because it has changed. Our people are no more living there, and nor are their spirits.' A few San still lived in the reserve in the old way—'San heroes', Omphile called them— and members of his family had returned to the reserve. But their numbers fell with each passing year.

Omphile took me to his family compound to sit with his aunt who was visiting from her home in the reserve. She came just once a year, at most twice, to visit her family and stock up on supplies. Omphile's aunt had no interest in talking with me and talked in generalities: Life in the reserve is not an easy life . . . there are problems with water . . . there are no doctors . . . life is simple out there . . . 'We are free,' she said, pronouncing the final word on the matter.

Back in Ghanzi after visiting New Xade, I had sat down with Kuela Kiema, the San author of *Tears for My Land* who had spent the first half of his life in the reserve. Kiema had long been a controversial figure among the San because he, like Scupa, had agreed with the government's decision to expel the San from the CKGR. For doing so, Kiema was, in his own words, denounced as 'a government puppet' and 'a sellout', and accused of having been bought off by the government.

But in Kiema's telling, the San had—as Scupa told me— betrayed their sacred pact with the land. By 1997 they were living settled lives in unsustainable CKGR settlements and engaged in commercial hunting—hunting on horseback and in 4WDs instead of on foot, hunting with rifles rather than bows and arrows—that would have emptied the reserve of its wildlife had it continued. 'The people were killing animals in large numbers,' Kiema had told me.

Unlike Scupa, however, Kiema understood that the choice facing his people in 1997 was a Faustian trade-off: to save the land they loved, they had to leave it. He felt his people's sorrow deeply, sorrow that arose in part from a decision he supported. In his book, he recalled his first visit back into the CKGR in 2000: 'When we got to Xade I looked in the direction of my traditional hut but, of course, it was not there. The area looked strange; the land was naked without the village. I felt emptiness in my soul. I felt desperate. I felt dead, demoralised, dispirited. I was helplessly hopeless. I closed my eyes. I had lost my land.'

But it was complicated. He understood, too, the call of the modern world—education, the internet, money. The best the San could hope for in their relationship with their land, he told me, was to 'commodify' their ancient knowledge, to build and lead a tourism industry in which, as cultural gatekeepers to the Kalahari, they could teach visitors the old ways. It was the only way, he feared, that traditional San culture could survive.

In the meantime, he worried about San youth who lived in New Xade. Three years ago, Kiema's nephew had set out on foot with two friends to hunt for small antelope close to New Xade. They soon became lost and argued over which way to go. They separated. The two friends made it back to New Xade, but Kiema's nephew did not. 'It was the rainy season, when there was plenty of tubers, everything. He spent two nights in the bush. He was desperate for water, but there were many sources of water—these holes that collect water and so on. Finally, he stumbled upon the Old Xade road, and the wildlife people found him, almost dead. He was lying on the road. He had no idea of what to eat, what to dig, and yet there was food everywhere.'

Now, back in Old Xade on my Kalahari crossing, I worried about the decision throughout the night. First I decided to continue south through the Kalahari and take my chances; then I succumbed to common sense and recognised the need to drive to New Xade and Ghanzi for fuel. Back and forth I continued throughout the night. I barely slept. By the time I rose I had decided, finally, to drive, tail between my legs, to Ghanzi and return later that day.

The day dawned cold and clear, a beautiful Kalahari morning. Disconsolate, I packed up my camp and set out on my way, driving slowly. At the first intersection of trails, I turned right, accelerating towards the south and away from the safety of Ghanzi.

I knew well this tendency in myself: the indecision, followed by a wise choice, then throwing caution to the wind when the moment of truth arrived. Past the deserted rangers' post I drove, hurrying beyond the turn-off to the remote salt pan of Gckwae, hoping all the while that momentum would triumph over wisdom. I kept going, window down and exhilarated. The first 20 kilometres were heavy going—the engine strained in low gear and the wheels spun in deep, soft sand. The fuel gauge continued its downward spiral, but I wasn't turning back now. I buried my sense of foreboding deep, hoping that I would make it through force of will and a stubborn refusal to yield.

I would drive as far as the Xaxa waterhole, where I would reassess my fuel situation. But by the time I arrived at the turn-off, I had gone too far, as I had known I would; it had all been a ruse to propel myself forward. Even if I retraced my steps to Xade Gate, I would still fall short of the nearest fuel station: I had passed the point of no return. There was nothing for it but to press on into the section of trail that had first drawn me to the Kalahari: the southern reaches of the reserve, the true homeland of the San before they were evicted in 1997, and where very few people now came. I set my course for Bape, a remote and uninhabited campsite 100 kilometres away to the south-east and deep into the Kalahari.

Almost immediately, as if to reward such brave disregard for good sense, the fuel gauge stabilised. At least I now knew I would make it to the campsite at Bape. I travelled through a cauterised land, a smouldering grassed landscape, with trees in places still tinged by the charcoal of recent fires. Animals were few, and the tracks were overgrown and little used; high-grass tussocks scraped the chassis as I passed.

Not long after the trail swung south, beyond the turn-off that led north-east to Molapo, I stopped, turned off the engine and stepped off the trail. The noise of the engine lived on for a few moments, ringing in my ears, and then was gone. I heard the wind, hot and blustery, which then died. Bathed in the purest silence I have known, deep in one of the largest deserts in Africa, I stood transfixed, dizzy even, at the enormity of it all—of where I was, of what I was doing and of what lay ahead.

And then I looked down. Before me in the sand, perhaps 10 metres from the trail, was a single human footprint, pointing south. It was barefoot, and perfectly formed, somewhere between an adult's and a child's foot in size. This was no tourist's footprint. No other travellers had passed this way in days, the mark was fresh, and the imprint lay well off the track. I looked around. What did I expect? I didn't know, but I did have the feeling that I was being watched. I saw nothing, no one. How long I stood there, I cannot say. An hour? Ten minutes? To this day I believe that I was not alone.

~

Bape is one of the silent places of the earth. Unlike other Kalahari campsites, birds were few. To the north, west and east, the Kalahari's grasslands rolled to a horizon broken only by low scrub and stunted trees. When the wind blew, the grasses hissed and the trees whispered; it sounded like approaching rain. But for most of the time, the silence was complete. I emptied my last jerry can of fuel into the tank. It was a solemn moment, and I will always associate the smell of petrol with that fraught moment when I wondered how far I would get before the tank ran dry. Otherwise, the Kalahari has no smell, save for the slight burn of dust in the nostrils.

The dead silence of midday. The long, torrid heat of afternoon. The softening light late in the day. The coming sound of night crickets. I lit a fire and thought of the San campfires of old.

The San would keep the fires burning throughout the night, for cooking and for warmth, to save everyone the considerable trouble of starting a new fire and to keep predators at bay; the San have a saying that 'the lion and the fire are never friends'. Throughout the night, someone would stir whenever the fire ebbed, adding kindling as needed. People would draw near to the fire to eat, wander off to talk with neighbours and then return. As the night wore on, the crowd around the fire invariably grew and told stories, discussed the day's events, talked about tomorrow. It was a form of entertainment, and a way to share and gather information. On a good night, old San stories would be recounted, with the more outgoing members among them leaping to their feet to act out much-loved folktales.

Ours is the first time in human history that these nightly gatherings no longer happen across much of the Kalahari. The Kalahari has not only fallen silent; its nights are darker too.

I realised then that I had not seen another human being throughout the whole day. Nor had I seen or heard lions since I arrived in the Kalahari. Unlike on my previous visits, when I had seen lions often and throughout the reserve, lions were noticeably absent from my journey and I could no longer ignore what was becoming self-evident: lions had become scarce in the Kalahari, where once they had been plentiful. As much as I embraced the quiet, I understood that, unlike other African wildernesses, this silence was not as it should be. In the Kalahari it was unnatural, a void where once there had been life. The Kalahari was one of the few places left in Africa where the presence of people—a people who lived in harmony with the land and its lions—had nurtured wilderness and protected its natural state. With the San gone and lions having disappeared from sight, the Kalahari was broken.

～

Perhaps two hours before sunrise, in the morning's cold sting, I woke enveloped in silence. When the wind came, it startled me; I was like the impala, suddenly alert, wondering if something had happened or was about to happen. I wondered if I should be afraid.

But there is a strange peace that comes from abandoning yourself to fate. Perhaps it is the release from the stress of decision-making that occurs when you have no option but to move forward. Whatever the reason, I had slept better than on any other night in the Kalahari.

On other mornings I was up and gone in no more than fifteen minutes, foregoing breakfast to make the most of morning light when animals were active. This morning, calm and in no hurry to learn my fate, I boiled myself a cup of tea and sat in the quietness of this quiet place. I had heard nothing during the night, and the light revealed no footprints, lion or otherwise, passing through camp. I had been truly alone out here.

With the sun not yet fully above the horizon, I drove away and rejoined the main track heading south. I knew that I still needed a miracle to arrive at the southern gate of the contiguous Khutse Game Reserve with fuel to spare. The best I could sensibly hope for was to limp into one of the campsites of Khutse and wait for help.

Just over 20 kilometres south of Bape, in a luminous patch of Kalahari apple-leaf and golden sands, first one, then two, then a handful of people came running out of the bush. They were San and they seemed like an apparition in the cold light of morning.

I suddenly remembered a story once told to me by a Tswana man. Many years ago, after the San had been expelled from the CKGR but before many had returned, he had crossed the central Kalahari from south to north with his girlfriend; they joined a convoy with two Dutch travellers. On the first night, they camped at 'an abandoned Bushman settlement', which we established was Mothomelo. 'In the morning, we woke and, as we talked, we realised that we had all had the same dream—that we were asleep, there at Mothomelo, and that for the whole night we were surrounded by a circle of Bushmen. They weren't aggressive or saying anything. They were just watching.' When they woke, the San were nowhere to be seen.

And yet, now, here they were.

Where was I? I asked.

Mothomelo came the reply

I had stumbled upon one of just three San settlements left in the CKGR. The people were desperate for food—I gave them sugar and pasta and tea. They dressed in modern clothing; one old man wore a heavy army greatcoat to ward off the cold. But they were back on the lands they had inhabited for tens of thousands of years.

How long have you lived here? I asked

We have always lived here.

How many are you?

We are few.

Where do you get your food?

What we find in the bush—wild berries, tubers.

Does the government bring you food?

No!

Do you hunt the wild animals?

We are too scared of the government.

But do you hunt sometimes?

. . . Sometimes . . .

What do you hunt?

Only those animals that we can eat.

And with that the villagers melted away, as quickly as they had come—young children, women, old men. The last survivors.

~

I had found the last remnants of the San, but where were the Kalahari's lions?

Not far beyond Mothomelo, trackside, two rusted 4WD wrecks lay abandoned, a reminder of the perils that stalk the foolhardy. A little further on, I emerged from a stand of light woodland into a sandy compound of four houses built of cane and thatch. Two apoplectic dogs launched at the vehicle from the shadows. This was Gugamma, and the place looked shuttered and abandoned, with no sign of life other than the crazed hounds. I would later learn that just one elderly lady lived there. 'Some people, they just don't want to leave,' one park official told me. 'She says her ancestors live there.'

I had almost made it through the CKGR. But as I crossed the Tropic of Capricorn and into Khutse Game Reserve, the southern extension of the CKGR, the petrol light came on. The dial was in the red.

In time, the grasslands and river valleys of the north yielded to the salt pans—Tshilwane, Mahurushele, Sekushuwe, Khwankwe—of the Kalahari's south. I camped alongside one such pan, at Moreswe, my final stop on the journey. As I prepared for my last night enveloped in the Kalahari's uncomplicated silences, I discovered that my final 5-litre container of water had leaked. When I opened the back door, water surged out, seeping away into the sand; its remains sloshed around on the floor. All I had left was a half-litre bottle of water. My fuel was gone. And where I camped was 40 kilometres from the nearest person. I fell into a troubled sleep.

~

When the lion roared the next morning, I cannot say for how long I stood silhouetted against the dawn sky. It felt like the longest moment of a soon-to-be-cut-short life, each breath suddenly precious.

A lion roars for many reasons: to announce its claims over a territory, to warn off rivals, to call to other members of its pride or coalition. In his epic study of the Serengeti's lions, Dr George Schaller described how when a lion roars, for whatever reason, it 'draws air deep into its chest, tightens its abdomen with great force to compress the air, and then releases it through its vocal cords, the sound erupting from the throat with such energy that it carries great distances'. Schaller wrote also of the 'almost ventriloquial quality of the sound'—my lion could have been anywhere. Until then I had always shared Schaller's joy upon hearing lions roar, but now I was more inclined to agree with Sir David Attenborough that the roars of lions 'are quite enough to chill the blood in the blackness of the night'.

Where the night-time roars had been a lion calling to his pride, or marking the boundaries of his territory, the lion's morning roar was something else altogether. 'The full roar', Schaller wrote, 'advertises the animal's presence. It denotes, "Here I am."'

Indeed.

In slow motion, panic rising, trembling, I turned my head. Otherwise I remained motionless. High grass around the campsite's perimeter, no more than 3 metres from where I stood, concealed the lion, and I waited, fully expecting it to erupt from behind the wall of grass at any moment.

Fatal attacks by lions on people in the Kalahari are rare, but Kalahari lore is littered with near misses. Not only did a lion leap over the sleeping Scupa, but it stalked Scupa's family the following day. Dabe's friend, the hunter, was attacked by a lion as he crouched behind a bush. George Silberbauer was told a story of how, in the 1960s, a lion attacked a hunting party over a 36-hour period, focusing relentlessly on one man. The man suffered terrible wounds but survived; the other men received not a scratch. As these thoughts flickered through my mind in the long seconds that I balanced on top of the vehicle, two competing ideas coalesced—a perverse desire to laugh at how absurd I must have appeared, and a fervent hope that my lion had nothing personal against me.

Minutes passed.

What to do? I thought of Craig Packer, who once encountered a lioness while he was on foot in the Serengeti; after a moment's pause—just a moment? Really?—he charged at the startled lion, clapping and shouting, and she ran off into the bush. But I didn't know where my lion was, and I wouldn't have had the courage to charge it if I did. If I climbed down off the vehicle and made for the cab, I would be ignoring the fact that I had no idea where the lion might be, underestimating a lion's speed and forgetting one of the most important lessons of safari life: for a lion, food runs. I remembered how, as an adolescent, I had been told not to run if, in the Australian bush, I came across a snake that was no further away from me than its own body length—that I would be safer standing still so that the snake would not sense my presence and feel itself cornered. That was all well and good, but when a snake crossed my path I ran like I had never run before.

Still not daring to move, high on my vehicle, I began to consider what it might mean to be killed and eaten by a lion. *Killed by a lion.* It sounded so innocent, this phrase that one writes so casually. Even as I thought of it, I imagined it happening to someone else, somewhere else.

Nothing moved. How long had I been holding my breath?

Once caught by a lion, its prey has little chance. 'While one lion grasps the throat,' Schaller wrote, 'others usually begin to eat, and the animal may die from loss of its blood and viscera rather than from strangulation and suffocation.

'After prey has been pulled down,' he continued, 'especially if this has been done quickly, it struggles surprisingly little, sometimes even failing to thrash its legs. An extreme instance of such behaviour was an uninjured buffalo that lay on its side while a lioness chewed on his tail. Animals in such situations seem to be in a state of shock.'

Or, as one passage puts it, a hopeless victim of a lion attack very often looks 'less the victim than the witness of its own execution'.

Implicit in all of this—although Schaller spares our sensitivities and does not point this out directly—is that many animals remain alive while lions feast on their innards. For one terrible moment I was chilled by the image of a lion tearing at my organs as I stared at the sky and thought frantic final thoughts, trying to fix my daughters' faces in my mind's eye for all eternity.

I breathed out slowly.

For want of a better idea and with slow, deliberate movements, I packed up the car. I gathered up the previous night's dishes. I stowed away the mattress. I fiddled with the latches. I have never felt more *aware* in my life, and when wind rustled the grasses that surrounded me, I became certain that it was the movement of an oncoming rush of tawny fur. I tensed in readiness, determined not to let my bowels empty in terror, hoping that I could die quickly and with dignity. But the lion, watching, remained hidden. I packed away my cooking utensils and stove. I stored my camp chair and mini ladder. The key to the ignition fell under the car. I crawled on all fours in the sand to retrieve it. Still the lion waited as I secured the vehicle, one latch at a time.

A metal clasp clattered against the aluminium, an unnatural din that rang out across the pan.

Finishing, finally, I climbed into the cab, and collapsed behind the wheel, breathing heavily. I don't know how long I sat there. I tried the key, and the engine stuttered into life—at least there was enough fuel to take me from here. Unable to bear the suspense any longer, tears flowing and hands trembling, I eased my vehicle out onto Moreswe Pan and drove away from the Kalahari, its lions still nowhere to be seen.

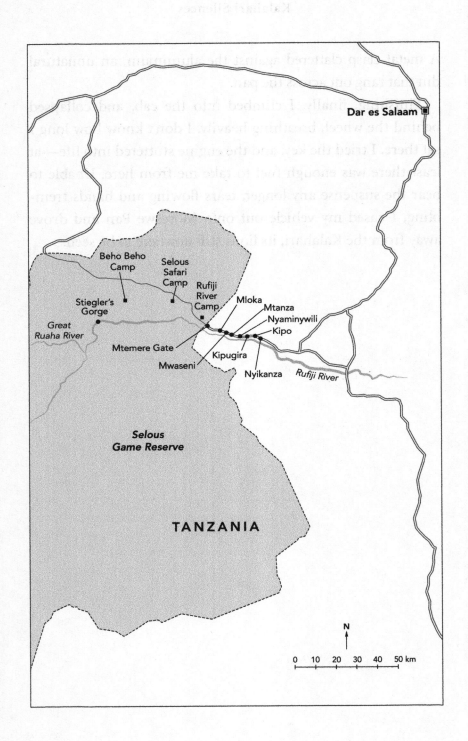

5

Among the Man-Eaters

Tanzania, 2019

The three lions came for her in the dark.[1] Perhaps they were hunting nearby and, having missed their usual prey, chanced upon Pili Amiri Namna as she climbed down the ladder to walk to the pit toilet a few metres away. Lions are expert nocturnal hunters, silent and stealthy; their night vision is far superior to that of human beings, who stumble blindly in the dark. She probably never saw them coming and perhaps never knew what hit her. Her husband and two-year-old son, asleep in the same shelter, awoke to her screams and lay terrified in the darkness, listening to the unmistakable sounds of lions eating their wife and mother alive.

It was the end of August 2002, the height of the dry season. Like most farmers in southern Tanzania, Pili Amiri Namna, in her early fifties, was sleeping in a shelter out in her fields: it was either that or lose her crops and risk starvation. August is when locals harvest maize, a subsistence crop. It's also when the local bush pigs—similar to warthogs, only marginally less ugly—feed on unguarded crops at night. Her village, Mwaseni, 240 kilometres by road south-west of Tanzania's largest city Dar es Salaam, runs like a ribbon along the northern bank of the Rufiji River. The river separates Mwaseni from its agricultural fields, a dangerous boat ride away. If she didn't spend the night in her fields, the bush pigs would feast. And anyway, she probably didn't mind: on the warm dry-season nights, her airy shelter, known as a *dungu*, was preferable, crop raiders or not, to the village's stifling four-walled structures; at least a cooling breeze was possible out in the fields and sleeping on the *dungu*'s platform a couple of metres above the ground.

The lions could have attacked anyone: there were plenty of candidates sleeping in similar shelters along the southern banks of the Rufiji that night. But it wasn't just anyone. It was a real person with a real family whose only crime was to be poor and to need to go to the toilet in the middle of the night. The next morning, all that remained of her was her skull and a few bones.

At first the attack seemed isolated, just wretched luck. But nearly two months later, just up the road in Kilimani, lions attacked and killed eighteen-year-old Musa Mutupu as he walked alone at night to his brother's *dungu*. No one heard his cries for help. If they had, perhaps another teenager, the son of a man named Mneka, would not have chosen to walk alone along a nearby path three nights later. He too was killed and eaten.

Over the next few weeks, killings by lions became almost commonplace in the area. Lions killed Hassan Gegedu, 62, in Ngorongoro, also along the Rufiji, and Said S. Matambwe, 48, in Kilimani, when they ran to help neighbours or relatives under attack from lions. Others were dragged from their shelters where they slept alone. On 14 November, lions in Kipugira took Said Mwenelwala, just seven years old, from his bed where he slept alongside his mother and three young siblings. On 22 December, lions killed Salum Mbonde, 80, in front of his three children in Mtanza. When his neighbour Yusuf Mbembe, 70, died three days later on Christmas Day 2002, killed when a lion grabbed his head in its jaws while his wife hung on to his leg in a macabre tug-of-war, lions had killed thirteen people in less than four months along the banks of the Rufiji River. Only one attack victim, twelve-year-old Hitanawe Kibana, survived.

At one point, villagers in Kilimani set a snare, captured a lion and killed it; some villagers ate the flesh of the fallen lion in the hope of gaining, if not strength and courage, then protection. But the killings continued. By the time that two lions took eleven-year-old Sadiki Dubuga as he walked home after eating dinner at his grandmother's hut in Kipugira in January 2003, it was clear that one of the most serious outbreaks of 'man-eating' lions in the modern era had gripped the villages along the Rufiji.[2]

For many, the damage was already done, and could never be undone. When Semeni Nasoro Malenda, Sadiki Dubuga's mother, learned of her son's death, she collapsed and never really recovered. Having lost her second child, this woman of lean beauty and quiet sadness would have no more children, a rarity in this land of large families. And when strangers turned up in her village sixteen years later to ask that she again tell the story of her loss, she passed out, so fresh and raw was her grief.

~

That lions should kill and eat people in modern Africa is less remarkable than that they don't do so more often. Lions don't normally eat people, because they don't see us as prey; they may view us as a threat, as a rival to be feared or vanquished, but not as food. Where lions do kill people, it is only because they have learned to do so. And when this happens, it is a catastrophe that looms as large over the history of our relationship with lions as it does over individual lives. When a lion kills a human, it scars the psyche and brings to mind all the night terrors and deepest fears of our kind. Lion expert Craig Packer is one of many to have argued that our fear of the dark may have begun with our ancestors' nocturnal encounters with lions and other predators.

The most famous outbreak of man-eating occurred in south-eastern Kenya in the dying days of the nineteenth century. As British colonial rulers raced to build their railway from the coast to the interior, work was halted in Tsavo—a wild, red-earthed land whose name means 'slaughter' in the local Kamba tongue—by two lions who preyed upon the Indian labourers hired to lay track. In one famous episode, the telegraph operator at tiny Kimaa station heard a lion leap onto the roof and begin tearing away at the corrugated iron. 'Lion fighting with station. Send urgent succour,' he tapped out to his superiors.

British officer Colonel John Henry Patterson, overseer of railway bridge construction on the Tsavo River, set out to kill the man-eaters, which he duly did; their rather diminished, taxidermied bodies now reside in Chicago's Field Museum of Natural History. In his picaresque version of the story, *The Man-Eaters of Tsavo*—and no doubt conscious of the power of exaggeration in securing his legacy as the Great White Bwana who saved countless lives through his bravery—Patterson claimed that the two lions killed 135 Indians and Africans. Later analyses suggested that the figure was closer to 35, but that was irrelevant to the power of the legend. No fewer than three movies about the Tsavo man-eaters have terrified audiences in the decades since: *Bwana Devil* (1952), *Killers of Kilimanjaro* (1959) and *The Ghost and the Darkness* (1996).

The story didn't end in Tsavo when Patterson killed the two offending lions in December 1898. In World War I, Tsavo became a theatre of war as Britain and Germany jostled for control over the colonies; Tsavo lay close to the frontier between British and German East Africa. One British officer, A.B. Percival, wrote how 'Lions were a veritable curse; man after man on sentry duty was taken, till it seemed sheer cruelty to put a man on such duty at all. Sentries were doubled and still men were taken. It takes little imagination to realise the case: two men at their post, bush all around and close-up—it was inviting attack by lions.'[3]

Although uncommon elsewhere, stories of man-eating by big cats echo down through the twentieth century with some frightening sets of numbers. Over five years in the 1920s, tigers were responsible for the deaths of 7000 people in India.

At around the same time, lions killed more than 161 people in the Sanga region of Uganda; one lion alone ate 84 people, while seventeen man-eating lions terrorised Uganda's Ankole district in the same years. More recently, there have been much smaller outbreaks around Mfuwe in Zambia in 1991, Queen Elizabeth National Park in Uganda in 1993 (where lions were unable to resist local men lying drunk by the roadside), and even in Tsavo in 1998. One outlier, according to Robert Frump in his *Man-Eaters of Eden*, is South Africa's Kruger National Park, where, he estimates, lions killed 13,380 refugees from Mozambique as they crossed the park between 1960 and 2005.[4]

Such special cases aside, if Tsavo was Africa's epicentre for man-eating lions in the late nineteenth and early twentieth centuries, that dubious honour shifted to Tanzania's south in the 1930s, and there it remains. No one knows how many people were eaten in the Njombe district south-west of Rufiji between 1932 and 1947, but it was somewhere between 1000 and 1500. In more recent years—and here we draw near to our current story—lions attacked over 1000 people in Tanzania from 1990 to 2007, killing around two-thirds of their victims. Seven districts in the south were responsible for more than half of all attacks. One of these districts was Rufiji.

Even when there is no 'outbreak', man-eaters are never far away in southern Tanzania: 'In a bad year, lions attack as many as 140 Tanzanians,' Craig Packer has written, and 'unreported cases may double that number. During quiet intervals, lions still attack ten to thirty people a year'. The survivor of one attack in Rufiji remembered how, when he was a child, fathers used to tell their children stories about dangerous lions, including one called Zimatar, The One Who Switches Off the Light. Man-eaters, it seems, have been a part of life here for generations.

It has long been the received wisdom that man-eating lions and tigers are old or injured cats and have turned to human beings because they are easier to catch and to eat. Studies of man-eaters have turned up evidence of tooth abscesses, and one of the Ugandan man-eaters in the 1990s had earlier been injured in a poacher's snare. Back in the 1950s, a Portuguese colonial official in Mozambique trapped a lion so that he could photograph it up close. The lion tore off its own foot trying to escape and, unable to hunt its customary prey, turned to killing people. One can only hope that his first victim was the said official.

The causes of man-eating are, more often, somewhat more prosaic. Some lions may have acquired a taste for human beings in centuries past when abandoned bodies littered the routes of slave caravans headed for the coast. Since the late nineteenth century, most outbreaks occurred when populations of lions' natural prey crashed, whether through game-control measures or rinderpest disease outbreaks, leaving lions hungry and desperate for a meal. In modern Tanzania, outbreaks of man-eating have occurred in areas where growing human populations have crowded out the preferred prey of lions. That lions can survive in such degraded landscapes owes everything to the presence of bush pigs. Unlike buffalo and impala, bush pigs love these human-dominated lands; they eat crops such as rice and maize, and they breed rapidly. With nothing else to eat, lions pursue the bush pigs into the fields, the same fields where farmers sleep in order to scare off the night-feeding pig pests.

That was what happened on the night of 31 August 2002, when three lions out hunting bush pigs came upon the unfortunate Pili Amiri Namna, across the river from Mwaseni. Ever the opportunists, they killed and ate her. Perhaps they liked the taste. Or maybe they enjoyed the ease of catching this new quarry. Whatever the reason, they taught their offspring how to do it. And the villagers along the banks of the Rufiji River were never the same again.

~

The killing of eleven-year-old Sadiki Dubuga in early January 2003 changed everything. Villagers fled their farms in terror, abandoning their crops. People stopped walking alone outside after dark. Parents gathered their children and the elderly indoors as darkness neared. The sense of a community under siege was hardly helped by this being one of the first large outbreaks of man-eating of the internet age: newspapers and online commentators from across the country reported each new death with rising hysteria, blaming a lion called Osama for the terror.

The lions became more brazen with each passing night. People were snatched from their beds, their palm-leaf walls no match for a lion's paw, and killings followed in Kipugira, Kipo and Nyikanza. No one knew where the lion would strike next.

Such are the ways of man-eaters. 'The movements of man-eating lions can be highly erratic and unexpected,' Packer has written. 'One famous man-eater in southern Tanzania was called Simba Karatasi, literally "paper lion", because he seemed to move about as randomly as a piece of paper blown by the wind.' The Rufiji lions would disappear for a few nights, kill a bush pig or two, and return miles up or down the river to strike again.

On 25 January 2003, a lion arrived at the door of Shamti 'China' Ngaona in Nyaminywili. As he retold the story sixteen years later, he smiled beatifically, his face creasing easily into a smile, as if he were recounting a benign tale of no great consequence and no little humour. Around him in the room of the village headquarters where he spoke, termite nests and mould stained the white walls, the dusty manila folders of the petty bureaucrat teetered on the shelves of a glass cabinet, and old calendars hung from the walls alongside windows without glass. He arrived alone and would later wander off alone, shuffling away across the bare concrete floor in his tattered runners, wearing an old Manchester United football shirt with the prominent lion of the Barclays Premier League on his sleeve patch. A loner, he had the smile of a simpleton, yet he was an enigma, and could be the village seer.

China recalled how he was on the ground floor of his *dungu* with his wife and three children. It was not long after dark when he opened the door to let out a stray cat that had entered looking for food scraps. With the door almost shut, a much larger feline threw itself against the door and a lion's paws broke through the gap—in letting out a stray cat, he had nearly let in a lion. More than the attack that followed, what chills the blood is the knowledge that a lion was stalking the house outside, preparing its attack, waiting for the right moment to tear these lives apart while China and his family, oblivious, went about their lives.

China and his wife threw their weight against the door. Although they managed to shut it, the lion swatted at the frame and began to break through the bamboo struts one by one. The lion's claws raked down China's groin; another claw strafed his forehead. The paw broke through again and China grabbed it, but it slipped from his grasp. China's children were, he remembered, very quiet as they watched the scene unfold at the entrance to a room no more than 6 or 7 square metres in size. They cowered in the corner as their lives hung in the balance. With the lion nearly through the door, China yelled at his wife for a spear. Suddenly, the lion ran off. Years later, China pointed to the timing of the lion's departure as evidence that this was not a real lion, but a spirit lion sent by malevolent people to harm them—'Why else did it run away when I asked for the spear? It understood what I said and knew for sure that I wanted to kill it!' And anyway, it must have been a spirit lion, he said, because a local government commissioner later told him, 'You have to do something to stop this. If you know that somebody is playing with you, you must do something.'

The screams of China and his wife alerted the neighbours, but if the people of Rufiji had learned anything it was that there was safety in numbers, and they waited for enough people to gather before running to help the family. All, that is, except 70-year-old Juma Makoma, who knew that if they arrived too late, an entire family might perish. Juma's two wives begged him to wait, then to let them go with him. He refused and insisted on running alone to the house of China and his family.

Even with all that had happened, the very idea of being eaten by a lion must have seemed so preposterous that he just assumed it could never happen to him. But having failed in its attack on the house, the lion turned its attention to Juma, the good Samaritan, and dragged him off into the night. Still too frightened to leave their house, China and his family listened to the unearthly screams of the man who had run bravely to their aid. Later the wailing of Makoma's two widows confirmed the outcome of the terrible night.

As China reached the end of his story, he was still smiling. But there was no longer any light in his eyes, and his face was a mask.

No one saw the lion again for two weeks, but it hadn't gone far. On 5 February, a lion attacked Mohamed Ngakoma, also in Nyaminywili. He survived, but only just: the lion's teeth clamped around the side of his head, causing a dislocated jaw and severe nerve damage; his face still sags to the left sixteen years later and he talks in a whisper. Otherwise, the lions went quiet. With each passing day, people began to wonder whether their ordeal might be over. It wasn't.

February turned into March, and with it came the rains. Within days, the land along the riverbank turned from dusty yellow to forest greens. March is when villagers plant rice in the newly flooded fields along the banks of the Rufiji. People returned to their farms, although whether this was because they felt safe or because they had little choice is impossible to say. Most are subsistence farmers, just one bad crop away from ruin, and bush pigs were everywhere—people just *had* to sleep in their fields to protect their most precious assets.

Abdallah Chembele, 33 at the time, was among them. On the evening of 9 March, he was in his *dungu* on the south bank of the Rufiji River, across the water from Kipo. With him was his mentally ill younger brother. Abdallah had always watched out for this brother, guiding him through life, protecting him from harm, whether self-inflicted or from others. He felt responsible. And so when this brother suddenly ran out into the darkness, Abdallah had little choice but to follow. It was around 7 p.m. and night had only fallen an hour before, but it was already a dark night, heavy with cloud. Not understanding the threat, the younger brother disappeared into the night— perhaps to him, it was a game of hide-and-seek. But Abdallah knew instantly that his brother was in mortal danger. He called to another brother in a nearby *dungu*, and they began to search frantically for their errant sibling. It was a desperate time, one in which their fears for their brother's wellbeing outweighed their own nervousness.

Long hours passed. Finally, close to midnight, they tracked their brother down. There was no time to waste. Relieved, angry and dead tired, they began the long walk back to the relative safety of their huts in the fields, walking in a tight line of three. There can be few terrors to match that of walking through the darkest of nights in the knowledge that man-eating lions are at large, and it began to mess with their minds: Abdallah felt that they were being followed. He stopped often to scan the surrounding bush with his torch, but he could see nothing. The fields and meandering pathways on the south side of the Rufiji are fringed with tall cashew trees, in whose shadows the darkness is even more intense, as well as dense vegetation

2–3 metres high, all capable of concealing a lion; in the rainy season, tall grasses hem everything in and it is easy to feel surrounded, even by day. As they neared their *dungus*, they passed through an area of high grass that each man lifted in order to pass. Abdallah was last in line, and something made him look over his shoulder. He could swear that he saw the grass move, as if a fourth person had passed behind them.

They stopped. Abdallah stared into the darkness. Nothing. Convinced that his mind was playing tricks on him, he laughed it off as another uncertainty of an already strange night. When his brother, sensing something, voiced his own fears that something or someone was following them, Abdallah dismissed the idea. He was afraid, but he knew that his brothers looked to him for reassurance. And anyway, they could see nothing; they had little choice but to continue.

They soon came to a slope. First one brother then the next headed down. Just as Abdallah began to descend, the lion made its move and leapt onto Abdallah from behind. Abdallah looked to his left and saw a lion's paw over his shoulder, to his right, another: Abdallah found himself staring into the lion's face. There are many metaphors for that moment—'staring into the lion's mouth' or 'looking death in the face' spring to mind. He remembered later that the lion, its mane and its maw, smelled rather like a male goat.

Abdallah was carrying an axe and, unable to move anything but his hands, he threw it towards his brothers, calling for help. They were too frozen with fear to move. It all happened in a split second, but Abdallah knew that the lion had misjudged its attack, probably because of the sloping terrain, and was now on its hind legs behind him. Whatever it was, Abdallah knew that the next few seconds were crucial: in those seconds he decided that he wanted to live.

The lion was using its back legs to try to knock Abdallah off balance, and its claw pierced his foot. Together they rolled down the slope, nearly breaking Abdallah's ankle with the weight of the fall, Abdallah imploring his brothers to act, and all the while trying to remain face down and beneath the lion. Lions strangle their prey, suffocating them at the throat. If the lion could shift its position, Abdallah knew that he would be dead in an instant.

Perhaps stung into action by Abdallah's struggle, his brothers finally came to his aid, shouting and charging the lion, which ran off. Abdallah sprang to his feet and felt suddenly cold; he was covered in blood. Until that moment, he had felt no pain. Adrenaline kicked in, and together the three brothers ran through the bush to the riverbank, followed by the angry growls of the lion that had missed its meal and was clearly neither happy about it nor far away. As they pushed out into the river's current in a wooden canoe, they looked back and saw the lion emerge from the bushes to watch them go.

Abdallah's physical wounds took six weeks to heal, and he then returned to his former life. Laconic, less scarred by the whole experience than bemused, he laughed off any sugges-tion that this was a spirit lion sent to harm him. 'I have been attacked by an elephant—it was chasing me, and I hid in a thorn bush. A hippo once attacked me, and I only escaped by climbing a tree. Another time when we were fishing on the lake, a crocodile attacked our canoe, and we hit it on the nose with the oars until it swam off.

'So when this happened with the lion, I thought it was just another time when I had to fight for my life. For ordinary people it is hard to imagine. But for me, because of the place I am living, these things are common.'

~

Mwaseni, the scene of the first attack back in August 2002, lies just 10 kilometres from the entrance to Selous Game Reserve, one of Africa's largest protected areas and home to what has long been considered the biggest population of lions on earth. Commonsense tells us that those two things—man-eating lion attacks, and a nearby, large and protected area filled with lions—are connected. The truth, as I learned, is far more complicated. But Selous casts a long shadow over events along the Rufiji River, and it is impossible to ignore.

For millennia, what we now know as the Selous was a thinly populated expanse of *miombo* woodland where impoverished communities eked out a meagre existence on the fringes of more prosperous lands to the east. Throughout history, as today, tsetse flies harassed humans and cattle alike, and served as a barrier to large-scale settlement. The tsetse may carry the scourge of African trypanosomiasis (sleeping sickness), but without it, conservationists have often argued, we would have very few national parks worthy of the name in Africa—having kept people at bay, and causing no harm to wild creatures, the tsetse kept large tracts of wilderness, including the Selous, from being settled. After Europeans set aside lands for hunting and for conservation, tsetse-fly wilderness areas were, in many cases, all that were left.

The Selous first entered European history books when the Scottish explorer Keith Johnston died in the Selous's Beho Beho Hills in 1879. He was with Joseph Thomson, whom we have already met in Kenya, on a Royal Geographical Society expedition to the lakes of Central Africa. Seventeen years later, a German governor of German East Africa announced a protected area on part of the site, and it became a hunting reserve in 1905. In his wonderful *Sand Rivers*, an account of a walking safari through a remote corner of the Selous in 1979, Peter Matthiessen recalls that Africans called the reserve Shamba wa Bibi (Wife's Land) after the reserve became 'a huge and shaggy present' to the wife of Kaiser Wilhelm II in the early twentieth century. During World War I, the British explorer, hunter and conservationist Frederick Courteney Selous was killed in a firefight in the reserve, but the British won the wider war; German East Africa was signed over to the British at the Paris Peace Conference in 1919, and the British announced the expansion of the reserve and named it after their hero three years later. It is not on record what the Kaiser's wife or her descendants—or any Africans, for that matter—thought of the whole affair.

By current standards, the reserve was a pitiful thing, barely 2500 square kilometres in scope, but it laid the foundation for what was to come. In the 1930s, one C.J.P. Ionides, game ranger for the colonial authorities, set about protecting the vast wilderness that is the Selous today. Under the pretext of eradicating sleeping sickness and protecting locals from wild animals, Ionides torched many of the isolated farms and villages, and added piecemeal to the reserve in the years that followed. Despite considerable local resistance, as sleeping sickness spread, so too did the boundaries of the reserve. Thus we owe this expansive African wilderness to what Matthiessen described as 'the organized depopulation of vast areas of southeast Tanzania'. The process continued into the 1970s, with the final expulsions taking place in 1974. A year later, the Selous Game Reserve took on its current form.

Descendants of those expelled from the reserve in the 1970s inhabit many of the villages along the Rufiji River where the lion attacks took place in 2002 and 2003. But resentment over the forced relocations has dulled with time, and many villagers declare themselves grateful that they no longer live in the wild Selous, and now have access to medical dispensaries and other infrastructure. 'Back in the old days,' according to Saidi Salumu Nyangalio, whose parents were among those forced out in 1974 and who settled in Kipo, 'your wife would go into the bush to look for firewood, and later you would hear that she had been taken by a wild animal.' Outbreaks of man-eating aside, he said, it is different now.

Selous Game Reserve now covers nearly 50,000 square kilometres, which makes it one of Africa's largest protected areas. But it is a strange kind of wilderness. Only 6 per cent (around 3000 square kilometres) is open for what is known as photo-tourism. The rest are hunting blocks or concessions, with their leases held by a handful of rich (and well-connected) Tanzanians and favoured foreigners. Even so, in lion conservation circles it has long been said that the Selous Game Reserve in its entirety is home to 4300 lions. *Four thousand three hundred.* Given that there are little more than 20,000 lions remaining on the continent, that would make the Selous home to 20 per cent of Africa's lions.

If only it were true. One lion researcher whose name I have withheld—lion numbers are a sensitive topic in Tanzania, which markets itself as Africa's last stronghold of the species— warned me against using the figure. No reserve-wide survey has ever been carried out and the widely used figure of 4300 lions in the Selous came from a survey of less than 1 per cent of the Selous's surface area.[5] More than that, the survey took place in what is probably the Selous's most favourable lion habitat and yielded a density of fourteen lions per 100 square kilo- metres, a figure that is plainly unreasonable for much of the rest of the park which consists of hunting concessions, many of them overhunted and poorly managed. For that reason, serious doubts exist that anywhere near 4300 lions live in the Selous.

And yet, even if the figure was closer to 1000 or even 2000, that's still a lot of lions. And the Selous is so large and so ecologically complete that its animals have, for the most part, no reason to step beyond the reserve's boundaries. We can argue over the exact figure, but the Selous remains one of Africa's most important lion populations.

~

March yielded to April. By the end of April 2003, lions had attacked 31 people in the Rufiji area since the previous August, and 23 of the victims were dead. Although it wasn't the best of rainy seasons, mist rose from the wet fields into the humid air and clouds hung low over the riverbank. Many people who were there at the time remember it as an oppressive period, a time of sweat and fear, of frayed nerves and wild rumours; it felt as if things were building towards something, although what that might be no one dared imagine.

All manner of opportunists flooded to the banks of the Rufiji, some to offer advice, others to prey on the villagers' fears. Primary among these were the spirit healers—or, as some locals call them, 'witchdoctors'—who roamed the villages; some were from the area, while others came from far away to try to extract profit from the midst of tragedy. Blaming spirit lions for the attacks, these witchdoctors promised intercession with the spirits and protection from the lions—in return, of course, for a small monetary consideration.

The issue divided the community, even families. On 30 April in Nyikanza, for example, 70-year-old Mussa S. Mbwate argued with his wife, Zaituni Omari Mbwate, 67, over this issue. Zaituni felt that it couldn't hurt to pay the 1000 Tanzanian shillings (around A$0.65). Mussa disagreed; he thought the witchdoctor a charlatan. It wasn't a real argument, and Mussa knew better than to force the issue: he had long ago learned when to let his wife have her way. Zaituni stood her ground and Mussa gave way without much of a fight. The witchdoctor's fee was a small but not inconsiderable sum for people in the region, and their son Asifiwi pointed out that the rains had been kind to them, the harvest was showing every sign of being a bumper crop, and the money wasn't much. In fact, just to stop their bickering, he, Asifiwi, announced that he would pay the fee on their behalf and be done with it.

Years later, Asifiwi, with his strong face, neatly trimmed beard and torn, grubby shirt, and smelling ever so slightly of alcohol, remembered how he had spent that afternoon with his father, mending fishing nets and laughing at the whole affair. Superstition was sweeping through Nyikanza and the other villages along the Rufiji at the time, and people were frightened, ready to try anything. Both father and son thought the witchdoctors were preying upon the gullible, and laughed at how most people hedged their bets and paid the spirit mediums in secret, even if they denied doing so. At the same time, Asifiwi told his father, why not pay up, just in case? It couldn't hurt.

Asifiwi and his parents went their separate ways—Asifiwi to his home, Mussa and Zaituni to sleep in the family *dungu* out in the fields. With them were two of their grandsons, Asifiwi's nephews, Mohamed Makonde, aged six, and Rajabu Watende, aged seven. Where they slept was almost like a tree house: a ladder climbed to a platform around 2 metres above the ground. It was the kind of adventure that young children along the Rufiji craved—a sleep-out with their grandparents, an afternoon of helping out with invented chores, an evening filled with games and stories and special treats. And when it came time to sleep, they would lie down in a neat row of four in the bare room with palm-leaf walls.

Around midnight, or perhaps it was later—either way, everyone was sound asleep—a male lion climbed the ladder, grabbed the sleeping grandmother Zaituni, and carried her down to the ground and out into the field. The lion began to feed on her, and her dying screams woke Mussa. Disoriented, the awful truth dawning, he cried out, calling for help to anyone who could hear them. Perhaps unaware until then that there were more people in the *dungu*, the lion stopped feeding, returned to the *dungu*, climbed the ladder and grabbed Mussa, who, we can imagine, was standing in brave defence of his grandchildren. The lion swatted him down, took Mussa in his jaws and carried him, too, out into the field.

What happened next is almost unbearable to imagine. One desperately wants to hope that Mohamed and Rajabu, who remained in the hut, slept through the entire ordeal. More likely, the two boys, aged six and seven, on their special night out with their grandparents, clung to each other for dear life, trembling in toe-curling fear as they wept and waited for the lion to return. And return the lion did. In no hurry, it climbed the ladder, seized one of the boys and took him off into the night. And then, in what must have seemed like an eternity for whichever boy was left, utterly alone in the world, the lion returned one final time to complete the family's obliteration.

The cries that tore at the fabric of the darkness must have been a haunting, terrible thing to hear. And the next morning, soon after daybreak, Asifiwi's seven-year-old son made the awful discovery of his bloodstained cousins and grandfather, and the dismembered body of his grandmother a short distance away. He ran back to his family. Drums began to sound across the village.

～

If the earlier killings had cast a pall of terror upon the communities of the Rufiji, the deaths of Asifiwi's parents and nephews—its scale, its awful malevolence—created a whole new level of fear. Events were spiralling out of control. Lions and other predators have been known to kill multiple prey—most often goats and other domestic livestock in fenced corrals—in a frenzy that has nothing to do with hunger. But to have returned so methodically and in silence for Mussa and the children after Zaituni had been taken had all the hallmarks, some villagers said, of a personal vendetta, of a creature so inured to any danger that it was killing for the sheer fun of it. Perhaps China had been right; perhaps this was a spirit lion, intent on inflicting maximum suffering upon this traumatised community.

There were many stories doing the rounds of what lay behind it all. One of the more widespread went like this. A fisherman came with his son from the Kilwa area (on Tanzania's coast) to fish in the Rufiji River. He caught many fish and the local people, becoming jealous and resorting to magic, sent a crocodile to kill the Kilwa man's son while he fished with his father. When the crocodile came for the son, it ignored the father, who sat at the back of the canoe, and went for the son at the front. As everybody knows, crocodiles always take the person at the back, so the father knew immediately what had happened. He called the villagers together and told them that he knew what they had done. Either they confess and give him back his son or, he warned, they would begin to lose the sons and daughters of their village. Then he returned to Kilwa, from where he sent the lion to take his revenge.

Many believed stories such as these. It mattered little that no one could actually remember such a visitor, or such a death, or such a village meeting. Perhaps it happened downstream. In any case, the idea of a spirit lion took hold.

And yet, others argued, hadn't Mussa and Zaituni paid the witchdoctor 1000 shillings precisely so that such a fate would not befall them? Among those who dismissed any notion that these were spirit lions was Juma Shamte Afa, the elected leader of Nyikanza village at the time. When news reached him early on the morning of 1 May that the four had been killed, he travelled to the site to verify the killings. Upon completion of this grisly task, he headed home to file a report for the authorities. Along the way, he met a group of recently arrived witchdoctors. Unaware of the killings, they petitioned him, as the village head, for an allowance or salary and for permission to take up residence in the village in order to protect it. When Afa pointed out that one among them had taken protection money from the Mbwates and that the Mbwates now lay dead, the delegation took his suggestion and left the village with great haste, never to return. Afa went further, calling on the villagers to take seriously this lion that was, he assured them, a real lion. He also argued that they should leave their farms and come into the safety of the village until the offending lion or lions were caught.

Many followed his advice, while others held fast to the notion of a lion sent from the spirit world to do them harm; the latter group believed that leaving their fields would offer no protection. Perhaps those who believed in the spirit lion felt grimly justified when, in a tragic twist of fate, Juma Shamte Afa lost his younger brother, Hatungani Afa, 25, to a lion attack six weeks after the killing of Asifiwi's family.

~

It should come as no surprise that the attacks on people by lions caused divisions, hysteria and a sense of helplessness within the communities of the Rufiji. As Craig Packer has written, 'It is difficult to exaggerate the toll that even a few man-eating lions can exact on the psychology of a rural community. Harvest season is man-eating season. Beyond the direct costs of injury and loss of life, people can become almost paralysed with fear, leaving their crops to rot in the fields.'

It wasn't as if the victims were doing anything out of the ordinary when they were attacked. Most were just going about their business. Some were going to the toilet. Others were sitting outside their huts or walking to their neighbours' shelters. Many, perhaps most were asleep in their beds when they were taken. One study of the man-eating outbreak in Rufiji after the fact by Dr Hadas Kushnir and Craig Packer found that 45 per cent of attacks occurred inside structures in agricultural fields, and 43 per cent of attacks occurred 'when individuals were resting, sitting, or sleeping inside their home'. There wasn't even safety in numbers: 59 per cent of victims were with other people when they were attacked. When the farmers of the Rufiji were asked where they felt most at risk of attack by lions, 43 per cent said it was in their fields, when they were farming and guarding crops—the cornerstone of their lives and livelihoods; 32 per cent felt that collecting firewood or building materials exposed them to danger from lions. It was, after all, an area where the simple task of fetching water could involve a two-hour walk through the bush.

This invasion by lions into the mundane heart of their lives sent terror through the communities. Each new story made its way up and down the river in no time at all, feeding the sense of fear that stalked even those communities where no attacks had taken place.

Partly this fear arose from the individual consequences of each attack, from scars that were more than physical. Mtoro Mohamedi Ngogi, for example, was attacked way back in 1998, more than four years before the 2002 outbreak, and his life has been forever marked by the experience. Just seven or eight years old at the time, he was asleep in his grandparents' *dungu* when a loud thump and the shaking of the structure woke him. He sat up, and although he was in between his grandparents, his sitting position made him the ideal target: the lion that had climbed the ladder swatted him so hard that he flew to the ground almost 2 metres below. It was only when he tried to stand that he saw and heard that it was a lion, a big male, which grunted—a low, menacing growl that suggested intent, and a sound every survivor of a lion attack is able to mimic. The lion came for him again and grasped Mtoro's head between its jaws. Mtoro's grandfather, woken by the noise, had the presence of mind to break off a palm leaf from the wall of the *dungu* and set fire to it, both for light and to scare off the lion. Neighbours gathered and, with his grandson being dragged by the head towards the bushes, Mtoro's grandfather implored them to help him confront the lion. Mtoro was still conscious; he remembered the pain of the lion's tooth piercing his skull, and, in deep contrast, the softness of the lion's mane that felt, he said, like a woman's wig. Some neighbours were too fearful to help but enough took up the cry to force the lion to drop Mtoro. The reprieve was only temporary. The lion charged again, and again clamped its jaws around Mtoro's head. The last thing he can remember is the searing pain before he blacked out.

His grandfather later told Mtoro that they had driven the lion off a second time and had rescued Mtoro, taking him to the village, then to Dar es Salaam where he spent a month in hospital. But he never saw again out of his right eye, which remains closed, encircled by a deep scar; whenever possible he wears a baseball cap tilted to that side. Once back in the village, he never returned to school—the loud noises made him feel like his head would explode, and other children teased him. His childhood had ended and with it any hope of ever completing his education. His options limited, barely able to read and write, he works as a day labourer on construction gangs when he can and helps out from time to time with the cleaning of tourist camps—but only when tourists are out on safari and he won't be seen.

More than two decades later, in a voice that always seemed on the verge of cracking, Mtoro remembered how 'I was under the complete control of that lion. No matter how hard I struggled, I couldn't break free'. He sighed, a sound that called to mind the harsh rasp of a lion's throat. More than anything else, he recalled, it was 'that feeling, that loss of control, that still disturbs me'.

~

The loss of control that Mtoro had felt soon extended out to the whole community around the Rufiji; the villagers felt they were no longer in control of their own lives. As Kushnir's research paper on the outbreaks put it, 'Only a small number of respondents accepted full personal responsibility for dealing with lion attacks . . . people feel somewhat detached from solutions; although lion attacks directly affect them, they do not feel like they have the ability to prevent future attacks.'

While that may have been understandable, there were some things that people could do. Some years after the outbreak, Packer and his team trialled a number of methods, from light sensors (to which bush pigs soon became accustomed) and noisemakers (that required constant maintenance) to fences built and ditches dug around agricultural fields. Fences and ditches were effective for a time. They may not have kept the lions out but they did prevent bush pigs from eating the crops; no bush pigs raiding crops meant that the farming families would not have to sleep in their fields but could instead sleep safely at home in their villages on the river's north bank. And even if they did decide to sleep in their fields, lions, which were still hunting bush pigs, would have less reason to enter the bush pig–free fields.

Theory was one thing, but practice was another altogether.

In the years after the trial, Nyikanza was diligent in maintaining the pilot project, its fences a model for what could be done if everybody in the community pulled together towards a common goal. The fences in Nyikanza lasted, with semi-regular maintenance, from 2009 until 2016, when an influx of herders changed everything along the Rufiji, and the fences went to ruin. One farmer, Shamti Hamisi Nyamlani (aka Mbulu), built fences under his own initiative and maintains them to this day. 'I have two wives,' he told me, with a twinkle in his eye. 'I don't want to leave my wives to go out and chase bush pigs.' Why don't other villagers follow suit? Don't they see that it works? 'It's like school. Everybody is taught the same thing, but not everyone learns the lesson.' Even so, the project in Nyikanza—and the fact that it lasted so long—was encouraging.

Not so in Kipo, where the fences crumbled and the ditches weathered away as soon as the researchers left. As one farmer said in 2019, 'you dug the trench when you came here for the project, and you maintained it for three years. But after three years, your project was done and we didn't see the need to do it anymore'. Did bush pigs still raid his crops? Yes, all the time. And were the lions still around? 'Yes, we heard them roaring just last night.' Another farmer blamed the influx of herders, but Kipo had given up on the ditches long before the herders came and tore down the fences in Nyikanza. 'Life is hard,' he said. 'The fences were expensive, and anyway, many people are too old to maintain them.'

No one could adequately explain why Nyikanza had maintained the fences and Kipo had not. But yes, life was hard for the villagers, and an outbreak of lions is an event of great power. How could they ever fight back against such a thing?

~

The loss of control that is at the heart of any encounter with a man-eating lion—the complete surrendering of power to the lion that Mtoro experienced—also influenced how people came to imagine their own risk of attack. After all, the rules of the game that prevail throughout Africa were not quite so clear in Rufiji, where lions now saw humans as prey.

The impact of this shift is easy to understand. As the research by Kushnir and Packer found, 96.5 per cent of villagers said that they were afraid of being attacked, 69 per cent were worried about being attacked and 53.2 per cent thought that they were very likely to be attacked. It didn't matter whether or not these people had been attacked by lions in the past. Instead, the study found, 'the disconnection between experience and risk perceptions is likely due to the extreme and uncontrollable nature of attacks . . . as well as the social amplification of risk, whereby discussion of attacks within the community may inflate concerns over the risk'.

The study went further, estimating that villagers in Rufiji District had a 0.19 per cent chance of being attacked by a lion during their lifetimes—higher, perhaps, than anywhere else on the planet but still a statistically tiny risk. And yet, most villagers ranked the likelihood of lion attacks as far higher. 'According to the United Nations World Food Program (2009), 58 per cent of Tanzania's population lives on less than

$1 a day, 44 per cent are undernourished, and 38 per cent of children under five are malnourished,' the study noted. 'The country is also plagued with irregular rainfall, and 1.4 million (3.4 per cent of the total population) are living with HIV/AIDS. Considering these statistics, it is remarkable that 40 per cent of the interviewees perceive the risk from lion attacks to be the same as drought, famine, malaria, and AIDS.'

Their explanation: 'Lion attacks mirror risks like terrorism or airplane crashes because even though attacks are unlikely, the consequences are high'—two-thirds of all lion attacks in the region resulted in death—'the situations are terrifying, and attacks are completely out of people's control.' Or to put it another way, 'the more sensational or vivid the consequences and the more the feeling of dread associated with the risk, the higher people perceive their own risk to be': people assume that something is more likely to occur if it comes easily to mind.

It requires very little imagination to sense how easily being eaten by a lion came to mind for the villagers of the Rufiji in 2003.

～

Studies of the man-eating outbreak came to one final and surprising conclusion: that living close to Selous Game Reserve, arguably the largest repository of lions on the planet, had little to do with the outbreak. Attacks occurred in villages along the Rufiji River (which runs due east away from the reserve), not along the park boundary, and distance from the reserve meant little in determining where attacks took place.

'We had expected to see a protected-area effect with attacks either being higher near sources of wildlife or higher in areas where lion prey is scarce,' reported Kushnir in her research paper. 'It is possible that resident lion populations in the agricultural areas are responsible for most incidents of man-eating, resulting in no clear link to protected areas.' In fact, more attacks by lions occurred in the coastal Lindi District between 1990 and 2007 than in Rufiji. Lindi has no national parks or protected lion populations of note.

This was an important finding. If living close to Selous Game Reserve had been pinpointed as a major risk factor when it came to man-eaters, then the implications for people who live close to national parks and other protected areas across Africa do not bear thinking about. The idea that national parks and their lions might represent a mortal danger to people living just beyond park boundaries could cause a political firestorm. It would also be a portent of a coming battle between lions and the burgeoning human populations that already surround Africa's last remaining protected areas.

You could almost hear governments and lion conservationists breathing a sigh of relief: Selous Game Reserve was not, after all, a threat to the people who lived close to its boundaries, and nor was it responsible for the man-eating outbreak that began in 2002. All well and good. But storm clouds gather over the Selous, and this could all change very soon.

~

The killings continued.

After the killing of the Mbatwe family at Nyikanza at the end of April 2003, lions killed two more people, one in Nyikanza, another up the road in Ngorongoro. Both were killed while sleeping in their *dungus* out in their fields. On 30 May, back in Nyikanza, exactly one month after the attack on the Mbwates, Siasa Hasani Afa, known by everyone as Jaba, was at home with his wife, Zaeni Kipalamoto, and their three children. It was around 6 p.m. and getting dark. Since the lion attacks had begun, the family had taken to eating their dinner early then going inside to bed before darkness fell. Once inside, they were safe—theirs was a sturdy home, not a flimsy *dungu*. On this night, everybody was inside, but Zaeni asked her husband to go outside with her so that she could wash her feet. Just beyond the door, she bent down to pour water on her feet, their eleven-month-old daughter tied to her back. Jaba had his back turned but heard a heavy sound, as if someone had fallen over. He turned and, in the failing light, saw a male lion on his wife's back, with the baby in between them; the lion was trying to manoeuvre itself so that it could grab Zaeni's throat with its jaws.

In tears, leaping to his feet to act out the scene, Jaba recalled how he grabbed a ladder from the family's chicken coop and tried to hit the lion, but it was impossible to strike a clean blow without also injuring his wife. The glancing blows had little effect upon the lion, who had Zaeni by the neck and was dragging her beneath its body towards the bushes; the baby was still trapped between them. From time to time the lion turned its head to see what was hitting it, but otherwise continued undeterred. Jaba was shouting and hitting the lion and knew that there was little time—the lion was close to disappearing with Zaeni and the child into the thick bushes where it would be impossible to follow. Only a deep ditch separated Zaeni and her child from oblivion, but when they entered the ditch, Jaba was finally able to get a clean hit on the lion. It dropped Zaeni and the child.

There are many heroes in these stories, and Jaba is one of them. The lion retreated a few steps, then turned to face them. Jaba put himself between his family and the lion and stood his ground, shouting, waving his ladder and trembling with fear. The lion charged again, pulling up just an arm's length from Jaba, who continued to hit at the lion. More than one life hung in the balance at that moment, and the stand-off could have gone either way. But Jaba sensed that the lion was afraid and, emboldened, he continued his defence of his fallen wife and child. When neighbours arrived on the scene, the lion slunk off into the bushes.

Jaba was sure that the lion would return, and when his badly injured wife called to him, 'Baba Hasan! Baba Hasan! I am here!'—in local culture, a wife addresses her husband as the father of their firstborn—he called for her to be quiet lest the lion hear her and return. At first the neighbours were too frightened to join him in the ditch. But when they saw Jaba pick up his wife and child, cradling them in his arms to carry them home, they rushed down to help. In an enduring testament to the sleeping powers of the African child, it was only as they reached the safety of home that the baby woke and began to cry.

Sadly, there are few happy endings when people are attacked by lions. Zaeni survived the initial attack but died from her injuries in hospital two days later. Their young daughter survived the attack on her mother, unscathed but for a few scratches, only to succumb to polio three years later. Jaba remains heartbroken.

It was nearly three weeks later, on 18 June, that Hatungani Afa, younger brother of Juma Shamte Afa (the village headman who had driven the witchdoctors away), was taken from a remote forest camp, plucked from a group of men sitting by the fire. He died instantly.

Then everything went quiet. The harvest was over, and many farms lay fallow. Although multiple lions had been involved in the attacks thus far, most accounts agree that by the end of the outbreak there was just one lion involved in the killings; after killing Hatungani Afa, it disappeared. Months passed. Many dared to hope that the lion had died or gone away, or had simply returned to eating its normal prey. But wherever the lion had gone in those months, it had not gone forever. On 19 August 2003, according to witnesses, it killed Halfani Mkumba, 38, at a fishing camp on the shores of Lake Kisiliwindu. His body was never found.

By November, people had again become complacent. It had been three months since the last attack and the dry season had arrived; almost all attacks had taken place during the rains and the time of harvests. On 6 November, Ally Sefu Mlimile was sitting outside his home on the south bank of the Rufiji, across the water from Nyaminywili. It was around 7 p.m., just dark. Ally had returned from a walk to a neighbour's house, his wife was cooking the family meal inside, and he was seated on the ground against a pole with his young son on his lap; they were playing together and listening to the radio. His parents were there, and they were talking, as families do, about nothing in particular.

Without warning, a male lion grabbed Ally over his right shoulder and clamped its jaws on the right side of his head. Ally's son was thrown clear, and the lion dragged Ally off by the head, pulling him beneath its body as it would a wildebeest or impala. Ally remembered fragments—the softness of the mane, the heat coming from the lion's body, and the realisation that, because of the pole against which Ally had been sitting, the lion hadn't quite gripped him as it would like. As he reached this point in his story sixteen years later, in a voice as deep as an elephant's belly rumble, he turned to reveal a scar that described a perfect arc, an inch above his right ear. Back on that November night in 2003, Ally cried out, 'Mother! I am being taken away!' Ally's father raised an almighty noise as he ran after the lion and his son. Having carried Ally around 10 metres, the lion dropped Ally and ran off. Ally, who was conscious throughout, stood, found that he was covered in blood and grabbed his father. Together they ran back to the house.

~

Selous Game Reserve is often marketed as one of Africa's last great wilderness areas: it is a vast tract of territory where no humans live, and wildlife populations are prolific.

And yet the Selous is in trouble in a way that could have far-reaching consequences for lions, for protected areas elsewhere in Tanzania and beyond, and, yes, for people living along the Rufiji River just outside the reserve's boundaries.

Let's begin with hunting. As we have already seen, a staggering 94 per cent of Selous Game Reserve consists of hunting concessions, places where safari operators fly rich (and mostly white) hunters into remote airstrips and hunting camps. Almost without exception, these hunters have paid tens of thousands of dollars for the right to hunt lions, leopards, elephants, buffaloes, kudus and other prized species. As we have seen in Zimbabwe, and while abhorrent to many, some conservationists argue that well-managed hunting blocks can ensure that landscapes and their overall wildlife populations are protected, and that poaching is prevented as the owners of the hunting leases safeguard their concessions for their wealthy clients. The only problem is that Tanzania's hunting industry remains shrouded in secrecy. Craig Packer, who ran the Serengeti Lion Project from 1978 until 2015, had his research permit revoked after mounting a campaign for greater transparency, accountability and sustainability in Tanzania's hunting industry. Packer's campaign cut a little close to home for the government: many of those holding leases for hunting concessions have close ties to those in power.

In those hunting areas for which Packer was able to obtain information, there were issues of profound concern. Lion populations were falling, often at frightening speed. This was largely because rather than shooting only males aged six and over—the then-minimum standard, backed by scientific evidence, for the sustainable shooting of male lions, although seven is now considered the minimum acceptable age—many hunters were shooting younger, sometimes even subadult males that had yet to grow a mane. Packer also found that other elements of a hunting operator's responsibilities—investment in wildlife management and anti-poaching measures, a willingness to open their books for scrutiny and verification, only pursuing sustainable quotas—were the exception rather than the rule. These failures, the paranoia and secrecy, the rampant corruption and cronyism in the awarding or renewal of hunting leases: Tanzania's hunting industry is rotten to the core, raising serious concerns about what may going on in the 94 per cent of the Selous controlled by hunters.

Trophy hunting is only part of the problem. UNESCO inscribed the Selous Game Reserve on its list of World Heritage Sites in 1982. This was a coup for the Tanzanian government, putting a reserve long considered one of Africa's best-kept secrets up there with global icons such as the Serengeti, Kruger and Yellowstone. Three years before, Peter Matthiessen's *Sand Rivers* had drawn attention to the plight of the reserve, but the threats to the reserve he described came more from government neglect and low-level poaching than anything else. If anything, Matthiessen's book served more to highlight the rustic charms of this rather remote and scruffy wilderness. But his expressions of concern over poaching and the government's intentions towards the Selous were an early warning that went unheeded. In recent years, those threats have grown, and grown rapidly.

In 2012, UNESCO's World Heritage Committee granted the Tanzanian government special permission to modify the boundary of the reserve, allowing it to excise around 200 square kilometres for the Mkuju River uranium mine in the far south. It sounded reasonable enough, until a 2017 report published by the WWF found that a further '34 mining concessions that overlap the Selous have been granted and a further 14 concessions have been applied for. There is active oil exploration in one overlapping concession where a sedimentary basin of interest overlaps the Selous . . . In general, there is practically no publicly available information on planned exploration or exploitation, and its impacts'. For all its bluster, the Tanzanian government has not disputed these figures.

Just two years after UNESCO approved the boundary modification, the Committee placed Selous Game Reserve on the World Heritage in Danger list, thanks to the very poaching crisis that Matthiessen and his travel companion, former game warden of the Selous Brian Nicholson, had foreshadowed in *Sand Rivers*.

There are many elements to poaching in the Selous, but it all boils down to this: when UNESCO recognised the Selous back in 1982, the reserve had close to 130,000 elephants; by 2013, there were just 13,000 left. According to the WWF, the poaching apocalypse—the 90 per cent crash in the elephant population—was caused by 'industrial-scale poaching', and there was little doubt that the crisis was accelerating. Even as late as 2006, there were still around 50,000 elephants in the Selous. By 2013, the WWF estimated, six elephants were being killed every day in the Selous, and 2190 elephants were being slaughtered every year.

And while 13,000 may still sound like a lot of elephants, studies have shown that the Selous could sustainably support an elephant population closer to 100,000. If you consider that there are only around 400,000 African elephants left on the continent, the fate of elephants in the Selous takes on a far greater significance.

~

The Selous was already under intolerable strain from trophy hunting, mining and poaching, but the government joined the assault on the reserve when it announced in 2017 that it was moving forward with a dam and hydroelectric power project at Stiegler's Gorge, in the heart of the Selous's photo-tourism safari area.

The idea of a dam in the Selous is as old as the reserve itself. An engineer named George Stiegler first floated the idea of a dam back in 1907. Soon after, he was killed by a charging elephant and his body fell into the gorge that would later carry his name. Over the century that followed, those who ruled what we now know as Tanzania paid lip-service to the project, but it was never more than talk. After the country gained independence in 1961, its new prime minister Julius Nyerere praised the idea of a dam as the 'potential foundation for the country's development and industrialisation'. He set up a commission to run the project in 1975, but momentum died and the project was shelved due to a lack of funding; even the World Bank, no enemy of dams, notably refused to bankroll its construction.

Things began to change in 2004, the latest in a number of drought years, when power cuts crippled the country's economy, particularly in Dar es Salaam. The national debate on the power crisis returned inevitably to the hydroelectric possibilities of a future dam in Stiegler's Gorge, and it has been in the news ever since. Funding remained the main obstacle— the estimated US$3.6 billion cost of the dam represents more than a quarter of the country's entire annual expenditure. But a 2010 visit to Tanzania by Brazilian president Luiz Inácio Lula da Silva, who was eager to expand his country's influence in Africa, was a game-changer. Brazilian company Odebrecht became heavily involved in the project and, in 2017, Tanzania's new president John Joseph Magufuli announced that construction would begin in 2019. When finished, the 126-metre-high dam across the 8-kilometre-long gorge would flood 1200 square kilometres of the reserve.

Undoubtedly construction of the dam spoke to a need within the country for a reliable power supply. In 2014, load-shedding (a scheduled power blackout in order to protect electricity reserves) was estimated to cost Tanzania up to 7 per cent of its GDP, and many Tanzanians, especially in rural areas, live in homes without electricity. If the government's promises are true, the dam will provide enough electricity to power the nation, with some even left over for export to neighbouring countries. 'If it's going to bring electricity, we need it,' said one safari camp manager. 'Many Tanzanians are living like animals. And it's for Tanzania's industrialisation. There will be job opportunities for me and my family.'

But numerous studies of the dam and its impact have

highlighted doubts over the dam's ability to fulfil its promises. The flow of water into the Rufiji River and its tributaries has fallen significantly over the past two decades, partly as a result of drought, but exacerbated by water offtakes, hundreds of miles upriver, for irrigation of government-backed dry-season rice cultivation. In some years, the rivers that are supposed to supply the dam run dry. And the 2017 WWF report found— without, it must be said, providing any details—that 'there are multiple other power supply options, including hydropower sites, with similar costs and lower risks'.

Condemnation of the decision to proceed with the dam was swift and withering. In 2018, UNESCO's World Heritage Committee added the dam to its official rationale for keeping Selous Game Reserve on the list of World Heritage in Danger. Speaking of its 'utmost concern', the Committee warned that construction of the dam carried 'a high likelihood of serious and irreversible damage to the Outstanding Universal Value' of the Selous. The International Union for the Conservation of Nature (IUCN), which had in the 1980s dismissed the dam's limited environmental impact, changed course in 2011 and called for the project to be abandoned before it had even begun. The IUCN's 2017 report called the dam 'fatally flawed', denounced the environmental impact assessment as a whitewash, and spoke of 'the potentially large, long-term and irreversible ecological and social impacts'. The IUCN's report also warned that 'Given the large footprint of the project in the heart of the Selous Game Reserve, the potential barrier effect from a 100km-long and 12km-wide reservoir, and the need to create and maintain access roads, supporting infrastructure and a permanent human presence, it is clear that the undisturbed and wild character of the area will be severely affected.'

The impacts of the dam extend far beyond the flooded area. Studies commissioned by the IUCN and WWF have suggested that changes in water flows downstream could negatively affect the livelihoods of more than 200,000 people who depend upon the Rufiji River. These include farmers who rely on the annual rainy-season flooding of the river, and fishing communities in the Rufiji Delta where the river empties into the Indian Ocean. Related possible implications include loss of critical water supplies through evaporation, the erosion of riverbanks and a loss of aquatic biodiversity. In a rare moment of candour, even the Tanzanian government, when nominating the Rufiji Delta wetlands as the Rufiji-Mafia-Kilwa Marine Ramsar Site in 2004, warned that the proposed dam could have 'severe impacts on the ecological balance downstream in the Rufiji Floodplain and Delta'. Noting the threats to both the Ramsar wetlands and the World Heritage–listed Selous Game Reserve, the 2017 WWF study observed that 'it is unprecedented to risk losing the integrity of not one, but two globally significant protected areas to a hydropower project.'[6]

Inside the reserve, the access roads to the construction site will almost certainly be used by poachers—the study commissioned by the WWF reported in 2017 that poachers were 'still using tracks created by oil exploration in the 1970s'. The roads and the dam itself will also result in the loss of habitat for a range of species, including black rhinos, African wild dogs, lions and elephants; there will be disruption from the noise and dust, and dangers caused by the increased movement of trucks and people through the reserve.

In the dry season of 2019, the main access road through the Matambwe photo-tourism area of the Selous was widened and upgraded. A quick survey of the traffic passing along this road in September revealed as many as fifteen construction vehicles passing in any given daylight hour. The dust and the noise were intolerable—many of the trees and thorn bushes that lined the road were permanently coated in orange dust. Experienced safari guides noted that the big herds of elephant, wildebeest and buffalo were a thing of the past. Where once lion sightings had been guaranteed, they said, it was becoming more and more difficult to find lions in the reserve. 'The numbers are very, very few. Some days we used to see twenty lions. Now you're lucky to see one,' said one guide. 'They've moved away because of the trucks and the noise and the dust,' said another.

There have been many threats to the Selous in the past, and each chipped away at the reserve's claims that it was protecting one of Africa's last great wilderness areas. But with the construction of the dam and the associated infrastructure, along with the influx of large numbers of labourers and vehicles, the Selous, it seemed, reached its tipping point in 2019. By anyone's standards, it could no longer be called a wilderness.

~

At the same time that the government was moving ahead with the Stiegler's Gorge dam in full public view, it was preparing the final nail in the Selous's coffin, far from view, deep in the heart of the reserve, on a massive scale and with no public debate. In May 2018, the government opened to tender six logging concessions in the middle of the reserve. Under the terms of the tender documents, the winning companies would be able to clear-fell an area covering 143,638.22 hectares (nearly 1500 square kilometres), containing 2,657,852 trees.

In 2019, at a ceremony launching construction at the dam site, President Magufuli further announced plans to turn the Matambwe photo-tourism sector of Selous Game Reserve, that 6 per cent portion, into Nyerere National Park. Strict protections would apply to the new national park. The remaining 94 per cent of the existing reserve, give or take a percentage point or two, would belong to the hunters and to the government and would disappear from view.

While all of this was happening, the NGOs tried their best to rally support, but none who operated within Tanzania were willing to go on record as being critical of the government. President Magufuli's administration has a record of revoking research permits for foreign nationals and NGOs, and of taking criticism of his policies as a personal affront. Or, as Craig Packer puts it, the Tanzanian government is 'militantly not going to be seen to be doing what they're told. They're like a nation of cats.'

Safari operators have been cowed into silence; several of them refused to be associated in any way with this book for fear of having their leases rescinded. The local NGOs are too scared to say anything. So here's something that no one else will say: if construction of the dam goes ahead, if the logging concessions are granted, the Selous Game Reserve, already under massive pressure from mining, poaching and unsustainable hunting, will become a protected area in name only. The Selous will have ceased to exist.

~

As the man-eating outbreak around the Rufiji River entered its third year in 2004, the man-eating lions had not so much lost the element of surprise as moved beyond any need for it; they did as they pleased.

No matter how hard the government tried, the lion was always one step ahead of those sent to kill it. Harunnah Lyimo, on his first real job out of wildlife college, was part of a team of eleven that included local game scouts, lion experts from the German Technical Cooperation Agency (GTZ), and officers from the Tanzanian government's wildlife division. They spent months at a time in the area, following up each report and tracking down the lions responsible for the deaths. When news came in they would rush to the site of the attack, then follow the tracks. If they didn't find the lion, they would set up a platform high in the trees from where they could shoot the lion if it passed.

It was frightening work, and the suspense was, at times, unbearable. Sometimes, Lyimo remembered, 'we had to follow the drag marks. And if you follow them properly, then there should be a lion where the drag marks end. That was scary. We would follow the marks through a tunnel of tall grass. The lion could be anywhere.' On one occasion they set a snare, and they knew they had their lion when its aggressive growls changed into pitiful whines of pain. The captured lion charged at the men when they drew near, but it died in a hail of bullets. Some lions were tracked after the event, others were caught devouring a human body. Nerves taut, the team endured long days when nothing seemed to be happening, then they would be called to suddenly undertake difficult and dangerous work. A pact arose between them: even if you suspected that it was your bullet that killed the lion, you never boasted nor claimed any glory for yourself.

Despite claims that one lion was responsible, the team knew that they were dealing with an entire pride. Lions are social animals. They share kills and participate together in hunts. An adult lion that kills will teach other adults to do likewise, and the adults will bring cubs to the carcass to feed. And besides, the team killed lions and caught them in the act, and still the killings continued. With each lion the team killed, they refused to declare victory, even as each one of them secretly hoped that the war had been won. By early 2004 the team had killed ten lions. Harunnah participated in seven of these killings. But the man-eating didn't stop.

At the end of January 2004, a lion killed a man near Kipo, returned for his wife but then left the couple's adult daughter unharmed. A couple of months later, with the rainy season in full swing, the lion returned to its most common strategy: killing a man sleeping alone in his *dungu* across the river from Nyikanza. Just two days later, in Ngorongoro, three fishermen were asleep on a sandbank in the middle of the Rufiji River. A male lion swam out to their camp and attacked 30-year-old Omary M. Matibwa. He was injured but his fellow fishermen scared the lion off by hitting it with burning pieces of wood. To escape, the lion swam across the river, but instead of swimming to the thinly populated south bank of the Rufiji, it crossed to the north.

Until that moment, the lions had remained on the south bank, able to take cover in the ample grass thickets that separated the fields. On the north side lay densely populated villages where people were unaccustomed to protecting themselves against lions. Here there were schools, and children were routinely left to play unattended. Bush pigs, the lions' main non-human prey, were much scarcer there. With a lion on the north bank, anything could happen.

For three days after crossing the river, the lion lay low, biding its time, getting hungrier with each passing day. Whether people on the north bank were aware that the lion was lurking nearby is unclear. Either way, just after 9 p.m. on 20 April 2004, the lion leapt onto the roof of a two-storey *dungu* and killed 70-year-old Salima S. Libaga and 60-year-old Asha S. Mlanzi, one after the other. Having killed its prey quickly, the lion dragged Mlanzi's body nearly 2 kilometres into the bush and ate half of it.

Then the lion made its fatal error.

Hyenas have evolved to eat what they can and then disappear quickly from a carcass; unless they are caught in the act, they rarely return to the scene of the crime. This becomes important when, for example, livestock herders lose a cow to predators and lace its carcass with poison in retaliation.

But lions will remain at the kill site, or not far away, returning to eat, often over the course of a week, until they've had their fill. This means that if you're looking for the lion that has killed something, it's unlikely to be far away. That was how some of the lions had been tracked down by the team and killed. And that was why, on the morning after Salima Libaga and Asha Mlanzi were killed, the team sent to track down the man-eaters could follow the lion's tracks, knowing that it would not have travelled far. Around 2 p.m., the team, accompanied by angry villagers, found the lion eating human remains. Confronted with an angry and larger-than-usual posse of pursuers, the lion fled. Everyone gave chase. As they closed in, some of the team climbed trees to get a clearer line of sight, while incensed villagers, no longer afraid, charged through the forest to flush the lion out. A villager fired the first shot. Wounded, the lion ran on, but then turned and charged as bullets and buckshot rained down upon him.

Harunnah Lyimo was on his way back from a supply trip to Dar es Salaam when the lion was killed. When he arrived, he congratulated the rest of the team, and they returned to camp, ready for news of the next attack.

April turned into May, and weeks passed. The rains ended; the harvest came in. Weeks turned into months. The end of the rains and of the harvest usually saw a dip in attacks as people returned to the villages; this time of year had been quiet before. After three months, the lion-tracking team packed up their camp and returned to their old jobs, in Dar es Salaam and elsewhere. As the year progressed, people went back to their daily lives, daring to hope that the nightmare was over. At first, members of the team found themselves checking their phones, half expecting an alert and a call to return to Rufiji. But the call never came.

Fifty-one people had been attacked. Thirty-five had died. And there it ended. The outbreak was over.

~

What do the troubles inside Selous Game Reserve have to do with man-eaters? Nothing. At least not yet.

What has happened in the Selous and along the banks of the Rufiji is so unique to southern Tanzania that it would be wrong to extrapolate the lessons of the Selous and the Rufiji to anywhere else in Africa. There are plenty of highly disrupted ecosystems across the continent—places with rapidly increasing human populations causing habitat destruction—where no lion has ever killed a person. In northern Tanzania, for example, researchers Bernard Kissui and Dennis Ikanda have observed that lions never attack the Maasai, except for those lying down drunk, despite lion populations living alongside people. Ikanda found that in northern Tanzania's Ngorongoro Conservation Area, 'a hungry lion will push a child aside to get to a good goat or cow'.[7] The same is true around Amboseli in southern Kenya.

Habitat loss has for more than a century been a major cause of wildlife disappearing from our planet, and history tells us that it is the wild animals, not people, that will pay the highest price for our destruction of wild places. Where human–wildlife conflict occurs without mitigation, without programs such as the Lion Guardians, Long Shields or African Parks to help lions and people to live together, lions will be wiped out, usually without killing a single person. The lion never wins.

If there is a lesson from the Selous for the rest of Africa, it remains an important one: where protected areas are under threat, whether from human population pressures on the boundary or through deliberate government policy, the results for lions and other wildlife can be catastrophic. That this should be possible in a region still considered by many to be a lion stronghold does not bode well for the future of Africa's lions. If lions are not safe even in places like Hwange National Park in Zimbabwe or Tanzania's Selous, are they safe anywhere?

But the Selous is—and has always been—different. Out there, with conditions getting worse, it is not difficult to imagine a scenario in which lions once again become man-eaters.

In the three months from November 2006, the Tanzanian government expelled more than 200,000 cattle from the Usangu wetlands, many miles upriver from the Rufiji and its tributaries, and with them tens of thousands of people from the herding communities of the Mang'ati, Maasai and Sukuma. These three groups, the Sukuma in particular, had long been blamed for all manner of terrible deeds, but their greatest crime, never proven, was that they and their cattle were responsible for reducing the flow of the Great Ruaha River, a critical tributary of the Rufiji. (It mattered little that, as we have already seen, government-backed dry-season rice cultivation schemes were most likely to blame for the river's decline.) The government didn't just kick the herders out: they physically relocated them. Nearly one-third of the herders, along with 65,636 livestock animals, were taken to Rufiji. Since then, their numbers have grown rapidly.

In an area with no history of livestock grazing, and in a classic African confrontation between settled farming communities and pastoralists, this massive and forced demographic shift has caused great tensions along the banks of the Rufiji. With considerable anecdotal evidence to support the claim, locals blame the herders for breaching and then dismantling many of the protective fences in Nyikanza that had survived from the pilot project in 2009 until 2016.

Tensions aside, the impact at a landscape level has been devastating. Although there are no statistics available, the grasslands and pockets of forest along the north bank of the Rufiji, which had previously escaped large-scale land clearing for agriculture, have disappeared. As Mbulu, the Nyikanza farmer with the last remaining fenced enclosure, observed, 'since the herders came, we no longer see elephant and buffalo. The cows eat all the grass. The herders clear the land. And there is no grass and no trees left for the wild animals.'

Human encroachment and habitat loss do not on their own cause outbreaks of man-eating, but they are always a factor when outbreaks occur. Studies have shown that rapid human population growth in Tanzania is at its most intense in areas closest to national parks. As forests around those protected areas are cleared to make way for human populations, livestock grazing and farming, ecosystems suffer. And Tanzania has lost more than one-third of its forest habitat since 1990.

In the short to medium term, these conditions can increase a community's vulnerability to lion attacks. After examining outbreaks of man-eating across southern Tanzania in the 1990s and into the 21st century, Craig Packer and his colleagues said as much: 'The total number of cases [of man-eating] has increased strikingly since 1990, probably because Tanzania's human population has risen . . . with an associated loss of lion prey outside the protected areas.'

It is southern Tanzania's population of bush pigs that sets the region apart—man-eating outbreaks simply don't happen elsewhere in Tanzania, in areas with no significant bush pig population. Here in Rufiji, bush pigs are the reason why outbreaks of man-eating are a near-constant peril of life in the area—without bush pigs, lions would not have enough prey to survive in these human-populated lands. And as long as there are bush pigs, the 2002–04 outbreak may not be unusual. From November 1991 until April 1998, for example, lions killed 58 people and wounded nineteen in Rufiji district. One of those nineteen survivors was Mtoro Mohamedi Ngogi, the child who could never return to school and whose life changed forever when he was attacked by a lion. So, too, into the future. Under such conditions of habitat degradation both inside and outside the reserve, lions may be forced to move into the community lands beyond the Selous. If that happens, man-eating may once again, as one study described it, 'become a viable option for lions'.

When that happens, we will again be faced with everybody's nightmare scenario, with the reminder that when things go wrong with lions, they go badly wrong—for lions and for people. All it will take is one displaced Selous lion out hunting bush pigs, missing its prey and happening upon some poor, unfortunate soul walking to the toilet on a night filled with darkness.

~

The end of the Rufiji outbreak didn't mean that lions no longer attacked people. It just meant that it happened less frequently.

Two years after the outbreak ended, in 2006, Yusuf Shabani Difika, 29, was attacked in an itinerant fishing camp close to the eastern boundary of Selous Game Reserve. The lion, which may have come from inside the reserve, grabbed Yusuf from behind, taking in its jaws first one and then both of his arms as he tried to defend himself. His fellow fishermen drove the lion away with burning firewood and rushed Yusuf to his home village of Mwaseni, 10 kilometres from the reserve gate and where the 2002 outbreak had begun, then to the district hospital. There doctors gave him a stark choice: keep your arms but risk a highly probable death from infection, or let them amputate both arms above the elbow. Perhaps if he had been a man of means, Yusuf could have travelled to a better hospital in Dar es Salaam, and there they might have been able to save his arms. Perhaps. But this was never an option, and he chose amputation.

A fisherman without arms is no fisherman at all and, now helpless, Yusuf sank into a deep depression. Unable to cope with a husband who could no longer provide for his family, Yusuf's wife left him, and he moved in with an uncle who cared for him. He had gone from being an able and able-bodied member of the community to one who had to rely on others for eating, bathing and other basic bodily needs. In 2011, *National Geographic* photographer Brent Stirton photographed Yusuf for a story, and used the image to tell part of Yusuf's story on Instagram three years later.

In August 2019, just weeks before my visit, Yusuf was crossing the river at night with a friend in a wooden dugout canoe. Their canoe was attacked by a hippo and capsized. The friend was able to swim to shore but Yusuf, without arms to help him swim, disappeared beneath the water. He never had a chance, and his body was never found. If he had simply drowned, his body would have washed up downstream. In all probability, Yusuf was eaten by a crocodile.

And so it was that the man who survived the lion was attacked by a hippopotamus and probably dispatched from his wretched life in the jaws of a crocodile. Such strange, grim battles to the death may be the facts of life along the banks of the Rufiji. They could also be the warning shots in a coming war, one in which there will be no winners.

Acknowledgements

This book could not have been written without Luke Hunter, formerly president of Panthera, now head of the Big Cats Program at the Wildlife Conservation Society (WCS). Luke introduced me to all of the major players in the lion world, making possible most of the stories in this book. He's also a friend, and a damned fine cat expert. Thanks mate.

I am fortunate to have spoken with both Paul Funston and Craig Packer, two of the world's premier lion scientists, at important moments throughout the research and writing of this book. Their wisdom and commonsense approach to conservation has taught me much about lions. Paul and Lise Hanssen were also generous with their hospitality in northern Namibia while en route between Liuwa and the Kalahari.

In Amboseli in Kenya, I shall be forever grateful to Meiteranga Kamunu Saitoti for telling me his story, and to Leela Hazzah and Stephanie Dolrenry for allowing me to spend so much time learning about their extraordinary work with the Lion Guardians. Thank you also to Philip Briggs and Eric Ole Kesoi, and to Darcy Ogada in Nairobi. Great Plains Conservation and Campi Ya Kanzi also generously provided accommodation for part of the time that I was in the Amboseli region.

In Zimbabwe, Andrew Loveridge was extremely generous in allowing unprecedented access to decades of data gathered at the Hwange Lion Research Project. On the ground in Hwange National Park, Jane Hunt was a wise and unfailingly patient teacher about the lives and individual stories of Hwange's lions. Brent Stapelkamp was also generous with his time and considerable wisdom about Hwange and its lions. Thanks also to Tendai Kedayi, a wonderful guide at Linkwasha and else-where, and to Calvet.

In Liuwa Plain National Park, special thanks to Jakob Tembo ('Mr Liuwa')—travel companion, lion expert and formidable Liuwa presence—and to Induna Mundandwe and the villagers of Kandiana. At African Parks, warm thanks to Rob Reid, Charlotte Pollard, Eva Meurs and Andrea Heydlauff, and to Daan Smit, Matthew Becker and Teddy Mulenga Mukula at the Zambian Carnivore Programme.

In and around the Central Kalahari Game Reserve, I am grateful to Dabe Sebitola, Kuela Kiema, Omphile Gabobonwe and Majwagana Tshuruu (Scupa) for recounting their lives to a complete stranger. Thank you also to Wilderness Safaris, here and in Hwange National Park in Zimbabwe, and to Kwando Safaris and Natural Selection, for allowing me to stay at their camps, something that would not have been possible without their generosity. Thanks also to Helena Fitchat, and to Andy Raggett of Drive Botswana.

In Tanzania, thank you to Harunnah Lyimo who was invalu-able as fixer, translator, local expert and travel companion in telling the man-eating aspect of the story; he also generously shared his research on the 2002–04 man-eating outbreak and connected me with the survivors. I am humbled by the trust placed in me by those who recounted their painful memories of man-eating lions in the villages along the banks of the Rufiji River: Semeni Nasoro Malenda, Shamti 'China' Ngaona, Mohamed Ngakoma, Abdallah Chembele, Asifiwi Mbwate, Juma Shamte Afa, Mtoro Mohamedi Ngogi, Siasa Hasani 'Jaba' Afa, and Ally Sefu Mlimile. Thanks, too, to Shamti Hamisi 'Mbulu' Nyamlani and Saidi Salumu Nyangalio. Numerous other people helped me in important ways in telling the story of the Selous Game Reserve, but I have chosen not to reveal their names so as to prevent any retribution against them from the Tanzanian government

For their encouragement at critical times, I am grateful to John Vaillant, David Quammen, Dereck Joubert, Hilary Rogers, Nick Lenaghan and Antoni Jach. My thanks also to Shivani Bhalla, Richard Bonham, Laurence Frank, Peter Ndirangu, Lisa Ham, Marina García and Alberto López de la Vega, Jane Leonard and Tobias McCorkell. My friendships with Damien De Bohun, Michael Dreelan, Claire McWalter, Matt Phillips and Philip Lee Harvey have long sustained me both personally and professionally. And to Alan Murphy, Wouter Vergeer and everyone at SafariBookings for their generosity in helping me get to Tanzania and Zimbabwe.

Allison Devereux, my agent at The Cheney Agency in New York, believed in this project from the beginning, and I shall be forever grateful for that belief. And to Grace Heifetz at Left Bank Literary, thanks, too, for taking this forward.

To everyone at Allen & Unwin, heartfelt thanks for getting behind this book. Publisher Jane Palfreyman saw the book's potential and carried it out into the world with great passion, wisdom and warmth. Emma Driver's edit made me a better writer and made this an infinitely better book through her clarity and attention to detail. Throughout the editing process, Samantha Kent was a reassuring presence, patient advisor and an unstinting champion of this project—Sam's wisdom, too, made this a better book.

And to my family, who have endured my long absences, and even longer periods when it appeared that these stories would never make it to a wider audience, I find words suddenly difficult to come by. To Jan, wise mother, faithful follower of my journeys, and one who sacrificed so many of her own dreams, you have earned the thanks of an eternally grateful son. And to my father, Ron: I wish you could have been here to read this book. To Marina, my companion along so many far-flung roads, I could not be more grateful. You kept the home fires burning and family life ticking over while I disappeared for weeks and months to follow this dream—*muchísimas gracias, amor mía.* And to Carlota and Valentina—who joined me on my journey to Liuwa and part of the Kalahari, and who are the most wonderful little people I can imagine—this book is for you.

Notes

Chapter 1

1. The dates used in chapter subheadings throughout this book refer to the date of my first visit to the country and region in question. Where I discuss historical events that took place before my visit, and later developments that occurred after I visited, I have made clear the dates of those events in the body of the narrative.

Chapter 2

1. The account of the Battle of Ngweshla that follows has been reconstructed based on the accounts of park rangers, lion researchers and safari guides.
2. This account draws heavily on what was told to me by Brent Stapelkamp. Stapelkamp was the public face of the Hwange Lion Research Project in the aftermath of what happened to Cecil, and four years later was still the world's leading expert on those events. He pieced together the story by talking with a close friend of Zane Bronkhorst. Bronkhorst is believed to have led the hunt. My telling of the story also draws on the account of Dr Andrew Loveridge.
3. A gin trap is a steel trap with serrated jaws or teeth designed to catch an animal by the leg or neck.
4. The common eland is the second-largest of all antelope species (after the giant eland)—it can reach a height of 1.78 metres at the shoulder, and can weigh up to 942 kilograms. Despite its size, an eland can reach speeds of 40 kilometres per hour and has great stamina, able to trot at 22 kilometres per hour almost indefinitely.

Chapter 4

1. G.B. Silberbauer, 1981, *Hunter & Habitat in the Central Kalahari Desert*, p. 12.
2. Quoted in J. Suzman, 2017, *Affluence without Abundance*, p. 48.
3. Quoted in J. Suzman, 2017, *Affluence without Abundance*, p. 51.
4. Quoted in J. Suzman, 2017, *Affluence without Abundance*, p. 65.

Chapter 5

1. The accounts in this chapter come from extensive interviews carried out in August–September 2019 with survivors of lion attacks or relatives of those who were killed. For those cases where I was unable to directly interview people involved in a particular case, I relied on unpublished interviews that were carried out by researcher Harunnah Lyimo in the aftermath of the attacks, and on discussions with him.
2. Lions do not discriminate when it comes to eating people, and eat man, woman and child as they find them, but 'man-eating' remains the widely accepted form of words.
3. Quoted in J.C.K. Peterhans & T.P. Gnoske, 2001, 'The science of "man-eating" among lions *Panthera leo* . . .', p. 12.
4. Frump's figure is quoted in A.J. Loveridge, 2018, *Lion Hearted*, p. 142.
5. This figure comes from P. Mésochina et al., 2010, *Conservation Status of the Lion* (Panthera leo *Linnaeus, 1758) in Tanzania*, and is based on the research of H. Brink et al., 2012, 'Methods for lion monitoring . . .'.
6. The Ramsar Convention on Wetlands of International Importance is an international treaty that recognises wetland areas for their importance to wildlife and broader ecosystems, and calls for their conservation and protection.
7. Quoted in C. Packer, 2009, 'Rational fear', p. 44.

Bibliography

Aduna, M.A. et al., 2018, 'Spatial and temporal trends of rainfall and temperature in the Amboseli ecosystem of Kenya', *World Journal of Innovative Research,* vol. 5, no. 5, pp. 28–42

African Development Bank, 2006, *Project Completion Report: Kapunga Rice Irrigation Project, Tanzania,* Abidjan: African Development Bank

African Parks, 2015, *Annual Report 2014: Our conservation model unpacked,* Johannesburg: African Parks

——2018, *Annual Report 2017: Restoration: Nature's return,* Johannesburg: African Parks

Baldus, R.D., 2004, 'Lion conservation in Tanzania leads to serious human–lion conflicts with a case study of a man-eating lion killing 35 people', Tanzania Wildlife Discussion Paper no. 41, GTZ Wildlife Programme in Tanzania

——2006, 'A man-eating lion (*Panthera leo*) from Tanzania with a toothache', *European Journal of Wildlife Research,* vol. 52, no. 1, pp. 59–62

Bauer, H., Chapron, G., Nowell, K., Henschel, P., Funston, P., Hunter, L.T.B., Macdonald, D.W. & Packer, C., 2015, 'Lion (*Panthera leo*) populations are declining rapidly across Africa, except in intensively managed areas', *Proceedings of the National Academy of Sciences,* vol. 112, no. 48, pp. 14894–9

Bauer, H., Chardonnet, P. & Nowell, K., 2005, 'Status and distribution of the lion (Panthera leo) in east and southern Africa', background paper for the East and Southern African Lion Conservation Workshop, Johannesburg, 11–13 January 2006

Bauer, H., Müller, L., Van der Goes, D. & Sillero-Zubiri, C., 2015, 'Financial compensation for damage to livestock by lions *Panthera leo* on community rangelands in Kenya', *Oryx*, vol. 51, no. 1, pp. 106–14

Bauer, H., Nowell, K., Breitenmoser, U., Jones, M. & Sillero-Zubiri, C., 2015, *Review of Lion Conservation Strategies: CMS working document*, Nairobi: UNEP

Bauer, H. & Van Der Merwe, W., 2004, 'Inventory of free-ranging lions *Panthera leo* in Africa', *Oryx*, vol. 38, no. 1, pp. 26–31

Becker, M.S., Watson, F.G.R., Droge, E., Leigh, K., Carlson, R.S. & Carlson, A.A., 2013, 'Estimating past and future male loss in three Zambian lion populations', *The Journal of Wildlife Management*, vol. 771, no. 1, pp. 128–42

Bertram, B., 1978, *Pride of Lions*, London: J.M. Dent & Sons

Beukes, B.O., Radloff, F.G.T. & Ferreira, S.M., 2017, 'Estimating African lion abundance in the southwestern Kgalagadi Transfrontier Park', *African Journal of Wildlife Research*, vol. 47, no. 1, pp. 10–23

Blackburn, S., Hopcraft, J.G.C., Ogutu, J.O., Matthiopoulos, J. & Frank, L., 2016, 'Human–wildlife conflict, benefit sharing and the survival of lions in pastoralist community-based conservancies', *Journal of Applied Ecology*, vol. 53, no. 4, pp. 1195–1205

Blumenberg, H. (trans. K. Driscoll), 2018, *Lions*, London: Seagull Books

Borrego, N. & Dowling, B., 2016, 'Lions (*Panthera leo*) solve, learn, and remember a novel resource acquisition problem', *Animal Cognition*, vol. 19, no. 5, pp. 1019–25

Borrego, N. & Gaines, M., 2016, 'Social carnivores outperform asocial carnivores on an innovative problem', *Animal Behaviour*, vol. 114, pp. 21–6

Briggs, H., 2019, 'Origin of modern humans "traced to Botswana"', *BBC News*, 28 October

Brink, H., Smith, R.J. & Skinner, K., 2012, 'Methods for lion monitoring: A comparison from the Selous Game Reserve, Tanzania', *African Journal of Ecology*, vol. 51, no. 2, pp. 366–75

Campbell, R., 2013, *The $200 Million Question: How much does trophy hunting really contribute to African communities?*, final report for The African Lion Coalition, prepared by Economists at Large, Melbourne

Caputo, P., 2002, *Ghosts of Tsavo: Stalking the mystery lions of East Africa*, Washington, D.C.: Adventure Press/National Geographic

Chan, E.K.F. et al., 2019, 'Human origins in a southern African palaeo-wetland and first migrations', *Nature*, vol. 575, no. 7781, pp. 185–9

Chardonnet, P., 2002, *Conservation of the African Lion: Contribution to a status survey*, Paris: International Foundation for the Conservation of Wildlife

Clarke, J., 2012, *Save Me from the Lion's Mouth: Exposing human-wildlife conflict in Africa*, Cape Town: Struik Nature

Cotterill, A., 1997, 'The economic viability of lions (*Panthera leo*) on a commercial wildlife ranch: Examples and management implications from a Zimbabwean case study', in J. van Heerden (ed.), *Proceedings of a Symposium on Lions and Leopards as Game Ranch Animals*, Onderstepoort, South Africa: South African Veterinary Association Wildlife Group, pp. 189–97

Creel, S. et al., 2013, 'Conserving large populations of lions—The argument for fences has holes', *Ecology Letters*, vol. 16, no. 11, p. 1413

Creel, S., M'soka, J., Dröge, E., Rosenblatt, E., Becker, M., Matandiko, W. & Simpamba, T., 2016, 'Assessing the sustainability of African lion trophy hunting, with recommendations for policy', *Ecological Applications*, vol. 26, no. 7, pp. 2347–57

Crosmary, W.-G. et al., 2018, 'Lion densities in Selous Game Reserve, Tanzania', *African Journal of Wildlife Research*, vol. 48, no. 1, pp. 1–6

Croze, H. & Lindsay, W.K., 2011, 'Amboseli ecosystem context', in H. Croze, C. Moss & P.C. Lee (eds), *The Amboseli Elephants: A long-term perspective on a long-lived mammal*, Chicago: University of Chicago Press, pp. 11–28

Cushman., S. et al., 2018, 'Prioritizing core areas, corridors and conflict hotspots for lion conservation in southern Africa', *PLoS ONE*, vol. 13, no. 7, e0196213

Daley, S., 1996, 'Botswana is pressing Bushmen to leave reserve', *The New York Times*, 14 July

Davidson, Z., Loveridge, A., Smith, K. & Macdonald, D., 2007, 'The value of lion as a component of the photographic/non-consumptive tourism market', poster presentation, Felid Biology and Conservation Conference, Oxford: University of Oxford, 18–20 September

Davidson, Z., Valeix, M., Loveridge, A.J., Madzikanda, H. & Macdonald, D.W., 2011, 'Socio-spatial behaviour of an African lion population following perturbation by sport hunting', *Biological Conservation*, vol. 144, no. 1, pp. 114–21

DeSantis, L.R.G. & Patterson, B.D., 2017, 'Dietary behaviour of man-eating lions as revealed by dental microwear textures', *Scientific Reports*, vol. 7, no. 907, pp. 1–7

Dickman, A., 2008, 'Key determinants of conflict between people and wildlife, particularly large carnivores, around Ruaha National Park, Tanzania', PhD thesis, London: University College London

Di Minin, E., Bradshaw, C. & Leader-Williams, N., 2016, 'Banning trophy hunting will exacerbate biodiversity loss', *Trends in Ecology & Evolution*, vol. 31, no. 2, pp. 99–102

Dolrenry, S., 2013, 'African lion (*Panthera leo*) behavior, monitoring, and survival in human-dominated landscapes', PhD thesis, Madison, WI: University of Wisconsin

Dolrenry, S., Hazzah, L. & Frank, L.G., 2016, 'Conservation and monitoring of a persecuted African lion population by Maasai warriors', *Conservation Biology*, vol. 30, no. 3, pp. 467–75

Dolrenry, S., Stenglein, J., Hazzah, L., Lutz, R.S. & Frank, L., 2014, 'A metapopulation approach to African lion (*Panthera leo*) conservation', *PLoS ONE*, vol. 9, no. 2, e88081

Dubas, C.L., 2016, 'The last roar of Africa's lions', *Africa Geographic*, 4 March

Dutton, E.A.T., 1929, *Kenya Mountain*, London: Jonathan Cape

Dye, B., 2017, *The Stiegler's Gorge Hydropower Dam Project: A briefing report for WWF*, Gland, Switzerland: WWF International

Dyson, P., 2015, 'Lion conservation in Maasailand: Response to Hazzah et al. 2014', *Conservation Biology*, vol. 29, no. 3, pp. 1–3

Elliot, N.B., Cushman, S.A., Macdonald, D.W. & Loveridge, A.J. 2014, 'The devil is in the dispersers: Predictions of landscape connectivity change with demography', *Journal of Applied Ecology*, vol. 51, no. 5, pp. 1169–78

Eloff, F.C., 1973, 'Water use by the Kalahari lion *Panthera leo vernayi*', *Koedoe*, vol. 16, no. 1, pp. 149–54

——1980, 'Cub mortality in the Kalahari lion *Panthera leo vernayi*', *Koedoe*, vol. 23, no. 1, pp. 163–70

——1984, 'Food ecology of the Kalahari lion *Panthera leo vernayi*', *Koedoe*, supplement, pp. 249–58

Estes, R.D., 1991 (2012), *The Behavior Guide to African Mammals*, Berkeley: University of California Press

Everatt, K.T., Moore, J.F. & Kerley, G.I.H., 2019, 'Africa's apex predator, the lion, is limited by interference and exploitative competition with humans', *Global Ecology & Conservation*, vol. 20

Ferreira, S.M., Govender, D. & Herbst, M., 2012, 'Conservation implications of Kalahari lion population dynamics', *African Journal of Ecology*, vol. 51, no. 1, pp. 176–9

Fihlani, P., 2014, 'Botswana Bushmen: Modern life is destroying us', *BBC News*, 7 January

Fitzherbert, E., Caro, T., Johnson, P.J., Macdonald, D.W. & Borgerhoff Mulder, M., 2014, 'From avengers to hunters: Leveraging collective action for the conservation of endangered lions', *Biological Conservation*, vol. 174, pp. 84–92

Foucault, M., 1966 (1994), *The Order of Things: An archaeology of the human sciences*, New York: Vintage Books

Frank, L., 2010, 'Hey presto! We made the lions disappear!', *SWARA—East African Wildlife Society*, vol. 4, pp. 16–21

——2011, 'Living with lions: Lessons from Laikipia', in N.J. Georgiadis (ed.), *Conserving Wildlife in African Landscapes: Kenya's Ewaso ecosystem*, Washington, D.C.: Smithsonian Institution Scholarly Press, pp. 73–84

Frank, L., Maclennan, S., Hazzah, L., Bonham, R. & Hill, T., 2006, 'Lion killing in the Amboseli-Tsavo ecosystem, 2001–2006, and its implications for Kenya's lion population', unpublished research paper

Frank, L. & Packer, C., 2003, 'Life without lions', *New Scientist*, letter, 25 October, p. 32

Frank, L., Hemson, G., Kushnir, H. & Packer, C., 2006, 'Lions, conflict and conservation in eastern and southern Africa', background paper for the East and Southern African Lion Conservation Workshop, Johannesburg, 11–13 January 2006

Frump, R.R., 2006, *The Man-Eaters of Eden: Life and death in Kruger National Park*, Lanham, MD: Lyons Press

Funston, P., Henschel, P., Hunter, L., Lindsey, P., Nowak, K., Vallianos, C. & Wood, K., 2017, *Beyond Cecil: Africa's lions in crisis*, New York/San Francisco: Panthera, WildAid & WildCRU

Funston, P.J., 2001, 'On the edge: Dying and living in the Kalahari', *Africa Geographic*, September, pp. 60–7

——2011, 'Population characteristics of lions (*Panthera leo*) in the Kgalagadi Transfrontier Park', *South African Journal of Wildlife Research*, vol. 41, no. 1, pp. 1–10

——2014, 'The Kavango-Zambezi Transfrontier Conservation Area—Critical for African lions', *Cat News*, vol. 60, pp. 4–7

——2018, *Securing Wildlife Populations of Hwange National Park, Zimbabwe—Panthera closure report to Fundación Loro Parque*, unpublished report

Funston, P.J., Groom, R.J. & Lindsey, P.A., 2013, 'Insights into the management of large carnivores for profitable wildlife-based land uses in African savannas', *PLoS ONE*, vol. 8, no. 3, e59044

Gilfillan, G., 2017, 'An investigation of the olfactory, vocal and multi-modal communication of African lions (*Panthera leo*) in the Okavango Delta, Botswana', PhD thesis, Brighton: University of Sussex

Government of Tanzania, Ministry of Natural Resources and Tourism, 2004, *Information Sheet on Ramsar Wetlands (RIS): Rufiji—Mafia—Kilwa Marine Ramsar Site*, 29 October, https://rsis.ramsar.org/RISapp/files/RISrep/TZ1443RIS.pdf, accessed 16 January 2019

Groom, R.J., Funston, P.J. & Mandisodza, R., 2014, 'Surveys of lions *Panthera leo* in protected areas in Zimbabwe yield disturbing results: What is driving the population collapse?', *Oryx*, vol. 48, no. 3, pp. 385–93

Guggisberg, C.A.W., 1961, *Simba: The life of the lion*, Cape Town: Howard Timms

Hanby, J. & Bygott, D., 1982, *Lions Share: The story of a Serengeti pride*, Boston: Houghton Mifflin

Hartmann, J., 2017, *The Stiegler's Gorge Hydropower Project: Rapid assessment of risks to the Selous World Heritage Site and the Rufiji-Mafia-Kilwa Marine Ramsar Site*, Gland, Switzerland: WWF International

Hazzah, L.N., 2006, 'Living among lions (*Panthera leo*): Coexistence or killing? Community attitudes towards conservation initiatives and the motivations behind lion killing in Kenyan Maasailand', PhD thesis, Madison, WI: University of Wisconsin

———2007, 'Living with lions in Kenya's Maasailand', *Wildlife Conservation Magazine*, March–April, pp. 6–9

Hazzah, L., Bath, A., Dolrenry, S., Dickman, A. & Frank, L., 2017, 'From attitudes to actions: Predictors of lion killing by Maasai warriors', *PLoS ONE*, vol. 12, no. 1, e0170796

Hazzah, L., Borgerhoff Mulder, M. & Frank, L., 2009, 'Lions and warriors: Social factors underlying declining African lion populations and the effect of incentive-based management in Kenya', *Biological Conservation*, vol. 142, no. 11, pp. 2428–37

Hazzah, L., Dolrenry, S., Kaplan, D. & Frank, L., 2013, 'The influence of park access during drought on attitudes toward wildlife and lion killing behaviour in Maasailand, Kenya', *Environmental Conservation*, vol. 40, no. 3, pp. 266–76

Hazzah, L., Dolrenry, S., Naughton, L., Edwards, C.T.T., Mwebi, O., Kearney, F. & Frank, L., 2014, 'Efficacy of two lion conservation programs in Maasailand, Kenya', *Conservation Biology*, vol. 28, no. 3, pp. 851–60

Hemson, G., 2003, 'The ecology and conservation of lions: Human–wildlife conflict in semi-arid Botswana', PhD thesis, Oxford: University of Oxford

Hemson, G., Maclennan, S., Mills, G., Johnson, P. & Macdonald, D., 2009, 'Community, lions, livestock and money: A spatial and social analysis of attitudes to wildlife and the conservation value of tourism in a human–carnivore conflict in Botswana', *Biological Conservation*, vol. 142, no. 11, pp. 2718–25

Henn, B.M. et al., 2011, 'Hunter-gatherer genomic diversity suggests a southern African origin for modern humans', *Proceedings of the National Academy of Sciences*, vol. 108, no. 13, pp. 5154–62

Humane Society of the United States (HSUS), 2015, *Trophy Madness: Elite hunters, animal trophies and Safari Club International's hunting awards*, September, Washington, D.C.: HSUS

Hunter, L., 2015, *Wild Cats of the World*, London: Bloomsbury

——2001, 'The future of Africa's magnificent cats', *Africa Geographic*, June, pp. 46–57

——2005, *Cats of Africa: Behavior, ecology, and conservation*, Baltimore: Johns Hopkins University Press

——2006, 'The vanishing lion', *Wildlife Conservation*, November/December, pp. 32–9

——2007, 'A battle royal', *Africa Geographic*, February, pp. 42–9

——2009, *Briefing Paper on Carbofuran*, Panthera

Hunter, L. & Barrett P., 2011, *A Field Guide to the Carnivores of the World*, London: New Holland

Hunter L.T.B., Pretorius, K., Carlisle, L.C., Rickelton, M., Walker, C., Slotow, R. & Skinner, J.D., 2007, 'Restoring lions *Panthera leo* to northern KwaZulu-Natal, South Africa: Short-term biological and technical success but equivocal long-term conservation', *Oryx*, vol. 41, no. 2, pp. 196–204

Ikanda, D. & Packer, C., 2008, 'Ritual vs. retaliatory killing of African lions in the Ngorongoro Conservation Area, Tanzania', *Endangered Species Research*, vol. 6, no. 1, pp. 67–74

International Union for the Conservation of Nature (IUCN), 2019, *Technical Review of the Environmental Impact Assessment for the Rufiji Hydropower Project in Selous Game Reserve, Tanzania*, April, Gland, Switzerland: IUCN

Isaacson, R., 2002, *The Healing Land: A Kalahari journey*, London: Fourth Estate

IUCN SSC Cat Specialist Group, 2006, *Conservation Strategy for the Lion in Eastern and Southern Africa*, December, Gland, Switzerland: IUCN

Jackson, D., 2010, *Lion*, London: Reaktion Books

Jeffers, H.P., 2002, *Roosevelt the Explorer: Teddy Roosevelt's amazing adventures as a naturalist, conservationist, and explorer*, New York: Taylor Trade Publishing

Joubert, D., n.d. 'Comparison of benefits between hunting and non-hunting use in Botswana', National Geographic Society, unpublished paper

Kays, R.W. & Patterson, B.D., 2002, 'Mane variation in African lions and its social correlates', *Canadian Journal of Zoology*, vol. 80, pp. 471–8

Kiema, K., 2010, *Tears for My Land: A social history of the Kua of the Central Kalahari Game Reserve, Tc'amnqoo*, Gaborone: Mmegi Publishing House

Kiishweko, O., 2013, 'Great Ruaha river that helps feed Tanzania under "alarming stress"', *The Guardian*, 16 January

Kingdon, J., *The Kingdon Field Guide to African Mammals*, 2nd edn, London: Bloomsbury

Kissui, B.M. & Packer, C., 2004, 'Top-down population regulation of a top predator: lions in the Ngorongoro Crater', *Proceedings of the Royal Society B: Biological Sciences*, vol. 271, no. 1550, pp. 1867–74

Kushnir, H., 2009, 'Lion attacks on humans in southeastern Tanzania: Risk factors and perceptions', PhD thesis, Minneapolis: University of Minnesota

Kushnir, H. & Packer, C., 2019, 'Perceptions of risk from man-eating lions in southeastern Tanzania', *Frontiers in Ecology and Evolution*, 28 February

Kushnir, H., Weisberg, S., Olson, E., Juntunen, T., Ikanda, D. & Packer, C., 2014, 'Using landscape characteristics to predict risk of lion attacks on humans in south-eastern Tanzania', *African Journal of Ecology*, vol. 52, pp. 524–32

Le Roux, W. & White, A., 2004, *Voices of the San*, Cape Town: Kwela Books

Lichtenfeld, L.L., 2005, 'Our shared kingdom at risk: Human–lion relationships in the 21st century', PhD thesis, New Haven, CT: Yale University

Lindsey, P.A. et al., 2018, 'More than \$1 billion needed annually to secure Africa's protected areas with lions', *Proceedings of the National Academy of Sciences*, vol. 115, no. 45, E10788–96

——2014, 'Underperformance of African protected area networks and the case for new conservation models: Insights from Zambia', *PLoS ONE*, vol. 14, no. 5, e94109

——2017, 'The performance of African protected areas for lions and their prey', *Biological Conservation*, vol. 209, pp. 137–49

Lindsey, P.A., Alexander, R., Frank, L.G., Mathieson, A. & Romañach, S.S., 2006, 'Potential of trophy hunting to create incentives for wildlife conservation in Africa where alternative wildlife-based land uses may not be viable', *Animal Conservation*, vol. 9, no. 2, pp. 283–91

Lindsey, P.A., Balme, G.A., Booth, V.R., Midlane, N., 2012, 'The significance of African lions for the financial viability of trophy hunting and the maintenance of wild land', *PLoS ONE*, vol. 7, no. 1, e29332

Lindsey, P.A., Balme, G., Henschel, P. & Hunter, L.T.B., 2016, 'Life after Cecil: Channelling global outrage into funding for conservation in Africa', *Conservation Letters*, vol. 9, no. 4, pp. 296–301

Lindsey, P.A., Balme, G.A., Funston, P., Henschel, P., Hunter, L., Madzikanda, H., Midlane, N. & Nyirenda, V., 2013, 'The trophy hunting of African lions: Scale, current management practices and factors undermining sustainability', *PLoS ONE*, vol. 8, no. 9, e73808

Lindsey, P.A., Roulet, P.A. & Romañach, S.S., 2007, 'Economic and conservation significance of the trophy hunting industry in sub-Saharan Africa', *Conservation Biology*, vol. 134, no. 4, pp. 455–69

Lion Guardians, 2008, *Annual Report 2008*, Nairobi: Lion Guardians
——2010, *Annual Report 2009*, Nairobi: Lion Guardians
——2011, *Annual Report 2010*, Nairobi: Lion Guardians
——2012, *Annual Report 2011*, Nairobi: Lion Guardians
——2013, *Annual Report 2012*, Nairobi: Lion Guardians
——2014, *Annual Report 2013*, Nairobi: Lion Guardians
——2015, *Annual Report 2014*, Nairobi: Lion Guardians
——2016, *Annual Report 2015*, Nairobi: Lion Guardians

Loveridge, A.J., 2018, *Lion Hearted: The life and death of Cecil & the future of Africa's iconic cats*, New York: Regan Arts

Loveridge, A.J. & Canney, S., 2009, 'African lion distribution and conservation modelling project—Final report', Horsham, UK: Born Free Foundation

Loveridge, A.J., Kuiper, T., Parry, R.H., Sibanda, L., Hunt, J.H., Stapelkamp, B., Sebele, L. & Macdonald, D.W., 2017, 'Bells, bomas and beefsteak: Complex patterns of human–predator conflict at the wildlife-agropastoral interface in Zimbabwe', *PeerJ*, vol. 5, e2898

Loveridge, A.J., Hemson, G., Davidson, Z. & Macdonald, D.W., 2010, 'African lions on the edge: Reserve boundaries as "attractive sinks"', in D.W. Macdonald & A. Loveridge (eds), *Biology and Conservation of Wild Felids*, Oxford: Oxford University Press, pp. 283–304

Loveridge, A.J., Hunt, J.E., Murindagomo, F. & Macdonald, D.W., 2006, 'Influence of drought on predation of elephant (*Loxodonta africana*) calves by lions (*Panthera leo*) in an African wooded savannah', *Journal of Zoology*, vol. 270, no. 3, pp. 523–30

Loveridge, A.J., Packer, C. & Dutton, A., 'Science and the recreational hunting of lions', in B. Dickson, J. Hutton & W.M. Adams (eds), *Recreational Hunting, Conservation and Rural Livelihoods*, Oxford: Blackwell Publishing Ltd & Zoological Society of London, pp. 108–24

Loveridge, A.J., Reynolds, J.C. & Milner-Gulland, E.J., 2007, 'Does sport hunting benefit conservation?', in D. Macdonald & K. Service (eds), *Key Topics in Conservation Biology*, Malden, MA: Wiley-Blackwell, pp. 222–40

Loveridge, A.J., Searle, A.W., Murindagomo, F. & Macdonald, D.W., 2007, 'The impact of sport-hunting on the population dynamics of an African lion population in a protected area', *Biological Conservation*, vol. 134, no. 4, pp. 548–58

Loveridge, A.J., Valeix, M., Chapron, G., Davidson, Z., Mtare, G. & Macdonald, D.W., 2016, 'Conservation of large predator populations: Demographic and spatial responses of African lions to the intensity of trophy hunting', *Biological Conservation*, vol. 204, part B, pp. 247–54

Loveridge, A.J., Valeix, M., Davidson, Z., Murindagomo, F., Fritz, H. & Macdonald, D.W., 2009, 'Changes in home range size of African lions in relation to pride size and prey biomass in a semi-arid savanna', *Ecography*, vol. 32, no. 6, pp. 953–62

Loveridge, A.J., Valeix, M., Elliot, N.B. & Macdonald, D.W., 2017, 'The landscape of anthropogenic mortality: How African lions respond to spatial variation in risk', *Journal of Applied Ecology*, vol. 54, no. 3, pp. 815–25

Lyimo, H., 2016, *Rufiji Field Revisit Assessment Report*, October (unpublished)

——*Rufiji Man-Eating Outbreak 2002–04: Chronology of Events* (unpublished)

Maclennan, S., 2006, 'Lions, livestock and spears', *Africa Geographic*, September, pp. 60–6

Maclennan, S.D., Groom, R.J., Macdonald, D.W. & Frank, L.G., 2009, 'Evaluation of a compensation scheme to bring about pastoralist tolerance of lions', *Biological Conservation*, vol. 142, no. 11, pp. 2419–27

MacSweeney, E., 2009, 'The lion saver', *Vogue*, November, pp. 236–41

Mallick, S. et al., 2016, 'The Simons Genome Diversity Project: 300 genomes from 142 diverse populations', *Nature*, vol. 536, no. 7624, pp. 201–6

Martin, G., 2012, *Game Changer: Animal rights and the fate of African wildlife*, Berkeley: University of California Press

Matthiessen, P., 1972, *The Tree Where Man Was Born*, London: The Harvill Press

——1981, *Sand Rivers*, Toronto: Bantam Books

McComb, K., Pusey, A., Packer, C. & Grinnell, J., 1993, 'Female lions can identify potentially infanticidal males from their roars', *Proceedings of the Royal Society of London*, vol. 252, pp. 59–64

McComb, K., Packer, C. & Pusey, A., 1994, 'Roaring and numerical assessment in contests between groups of female lions, *Panthera leo*', *Animal Behaviour*, vol. 47, no. 2, pp. 379–87

Mésochina, P., Mbangwa, O., Chardonnet, P., Mosha, R., Mtui, B., Drouet, N., Crosmary, W. & Kissui, B., 2010, *Conservation Status of the Lion (*Panthera leo *Linnaeus, 1758) in Tanzania*, Paris: SCI Foundation, DNPW & IGF Foundation

Miller, J.R.B., 2016, 'Aging traits and sustainable trophy hunting of African lions', *Biological Conservation*, vol. 201, pp. 160–68

Mills, G. & Mills, M., 2010, *Hyena Nights & Kalahari Days*, Auckland Park, South Africa: Jacana Media

Ministry of Natural Resources and Tourism, Tanzania Forest Services (TFS) Agency, 2018, *Tender No. AE-068/2017–2018/HQ/D/01 for the Sale of Standing Trees with a Total Volume of 3,495,362.823 M³ at Rufiji District—Tendering Document*, 25 April, Dar es Salaam: TFS

——2018, *Additional Information on Tree Species*, 8 May, Dar es Salaam: TFS

Mogensen, N.L., Ogutu, J.O. & Dabelsteen, T., 2011, 'The effects of pastoralism and protection on lion behaviour, demography and space use in the Mara Region of Kenya', *African Zoology*, vol. 46, no. 1, pp. 78–87

Monbiot, G., 1984 (2003), *No Man's Land: An investigative journey through Kenya and Tanzania*, London: Green Books

Morandin, C., Loveridge, A.J., Segelbacher, G., Elliot, N., Madzikanda, H., Macdonald, D.W. & Hoeglund, J., 2014, 'Gene flow and immigration: Genetic diversity and population structure of lions (*Panthera leo*) in Hwange National Park, Zimbabwe', *Conservation Genetics*, vol. 15, no. 3, pp. 697–706

Moss, C., 2000, *Elephant Memories: Thirteen years in the life of an elephant family*, Chicago: University of Chicago Press

Mosser, A. & Packer, C., 2009, 'Group territoriality and the benefits of sociality in the African lion, *Panthera leo*', *Animal Behaviour*, vol. 78, no. 2, pp. 359–70

M'soka, J., Creel, S., Becker, M.S. & Droge, E., 2016, 'Spotted hyaena survival and density in a lion depleted ecosystem: The effects of prey availability, humans and competition between large carnivores in African savannahs', *Biological Conservation*, vol. 201, pp. 348–55

National Assembly of Kenya, *Official Report—Parliamentary Debates*, 2 June 2009

Nowell, K., Hunter, L. & Bauer, H., 2006, 'African lion conservation strategies', *Cat News*, vol. 44, p. 14

Odino, M. & Ogada, D.L., 2008, *Furadan Use in Kenya and its Impacts on Birds and Other Wildlife: A survey of the regulatory agency, distributors and end-users of this highly toxic pesticide*, January, Nairobi: National Museums of Kenya

Ogada, M.O., Woodroffe, R., Oguge, N.O. & Frank, L.G., 2003, 'Limiting depredation by African carnivores: The role of livestock husbandry', *Conservation Biology*, vol. 17, no. 6, pp. 1521–30

Ogutu, J., Owen-Smith, N., Piepho, H.P. & Said, M.Y., 2011, 'Continuing wildlife population declines and range contraction in the Mara region of Kenya during 1977–2009', *Journal of Zoology*, vol. 285, no. 2, pp. 99–109

Oriol-Cotterill, A., Macdonald, D.W., Valeix, M., Ekwanga, S. & Frank, L.G., 2015, 'Spatiotemporal patterns of lion space use in a human-dominated landscape', *Animal Behaviour*, vol. 101, pp. 27–39

Owens, M. & Owens, D., 1984 (1994), *Cry of the Kalahari*, London: HarperCollins

Packer, C., 1994, *Into Africa*, Chicago: University of Chicago Press

——2009, 'Rational fear', *Natural History Magazine*, May, pp. 43–7

——2010, 'The economics of trophy hunting in Africa', *SWARA—East African Wildlife Society*, vol. 4, pp. 1–4

——2015, *Lions in the Balance: Man-eaters, manes, and men with guns*, Chicago: University of Chicago Press

Packer, C. et al., 2009, 'Sport hunting, predator control and conservation of large carnivores', *PLoS ONE*, vol. 4, no. 6, e5941

——2013, 'Conserving large carnivores: Dollars and fence', *Ecology Letters*, vol. 16, no. 5, pp. 635–41

——2013, 'The case for fencing remains intact', *Ecology Letters*, vol. 16, no. 11, pp. 1414–e4

Packer, C., Brink, H., Kissui, B.M., Maliti, H., Kushnir, H. & Caro, T., 2010, 'Effects of trophy hunting on lion and leopard populations in Tanzania', *Conservation Biology*, vol. 25, no. 1, pp. 142–53

Packer, C., Ikanda, D., Kissui, B. & Kushnir, H., 2005, 'Lion attacks on humans in Tanzania', *Nature*, vol. 436, no. 7053, pp. 927–8

Packer, C., Scheel, D. & Pusey, A.E., 1990, 'Why lions form groups: Food is not enough', *American Naturalist*, vol. 136, no. 1, pp. 1–19

Packer, C., Swanson, A., Ikanda, D. & Kushnir, H., 2011, 'Fear of darkness, the full moon and the nocturnal ecology of African lions', *PLoS ONE*, vol. 6, no. 7, e22285

Panthera, 2016, *Annual Report 2015*, New York: Panthera

Patterson, B.D., 2005, 'Living with lions—The notorious Tsavo lions', *Travel News and Lifestyle*, vol. 129, pp. 28–31

Patterson, B.D., Kasiki, S.M., Selempo, E. & Kays, R.W., 2004, 'Livestock predation by lions (*Panthera leo*) and other carnivores on ranches neighboring Tsavo National Parks, Kenya', *Biological Conservation*, vol. 119, no. 4, pp. 507–16

Patterson, B.D., Kays, R.W., Kasiki, S.M. & Sebestyen, V.M., 2006, 'Developmental effects of climate on the lion's mane (*Panthera leo*)', *Journal of Mammalogy*, vol. 87, no. 2, pp. 193–200

Patterson, J.H., 1907, *The Man-Eaters of Tsavo and Other East African Adventures*, London: Macmillan

Percival, A.B., 1928, *A Game Ranger on Safari*, London: Whitefriars Press

Périquet, S., Mapendere, C., Revilla, E., Banda, J., Macdonald, D.W., Loveridge, A.J. & Fritz, H., 2016, 'A potential role for interference competition with lions in den selection and attendance by spotted hyaenas', *Mammalian Biology*, vol. 81, no. 3, pp. 227–34

Peterhans, J.C.K. & Gnoske, T.P., 2001, 'The science of "man-eating" among lions *Panthera leo* with a reconstruction of the natural history of the "man-eaters of Tsavo"', *Journal of East African Natural History*, vol. 90, no. 1, pp. 1–40

Power, R.J. & Compion, R.X.S., 2009, 'Lion predation on elephants in the Savuti, Chobe National Park, Botswana', *African Zoology*, vol. 44, no. 1, pp. 36–44

Pusey, A.E. & Packer, C., 1994, 'Infanticide in lions: Consequences and counterstrategies', in S. Parmigiani & F.S. vom Saal (eds), *Infanticide and Parental Care*, Amsterdam: Taylor and Francis, pp. 277–99

Quammen, D., 2003, *Monster of God: The man-eating predator in the jungles of history and the mind*, New York: W.W. Norton & Company

Queeny, E.M., 1954, 'Spearing lions with Africa's Maasai', *National Geographic*, October, pp. 487–517

Ramsauer, S.N., 2006, 'Living at low density: A study of within and between pride dynamics in Kalahari lions', PhD thesis, Zürich: Universität Zürich

Reid, R., 2017, 'Remembering Lady Liuwa', *African Parks*, 10 August

Reid, R.S. et al., 2008, 'Fragmentation of a peri-urban savanna, Athi-Kaputiei Plains', in K.A. Galvin, R.S. Reid, R.H. Behnke Jr & N.T. Hobbs (eds), *Fragmentation in Semi-Arid and Arid Landscapes: Consequences for human and natural systems*, Dordrecht: Springer, pp. 195–224

Richards, N., 2011, *Carbofuran and Wildlife Poisoning: Global perspectives and forensic approaches*, Chichester, UK: Wiley

——2012, 'The carbofuran controversy', *Chemistry & Industry*, vol. 76, no. 2, pp. 22–5

Riggio, J.S., 'The African lion (*Panthera leo leo*): A continent-wide species distribution study and population analysis', master's thesis, Durham: Duke University

Riggio, J. et al., 2012, 'The size of savannah Africa: a lion's (*Panthera leo*) view', *Biodiversity & Conservation*, vol. 22, pp. 1–19

Ripple, W.J. & Beschta, R.L., 2012, 'Trophic cascades in Yellowstone: The first 15 years after wolf reintroduction', *Biological Conservation*, vol. 145, pp.205–13

Rodgers, W.A., 1974, 'The lion (*Panthera leo*, Linn.) population of the eastern Selous Game Reserve', *East African Wildlife Journal*, vol. 12, no. 4, pp. 313–17

Schaller, G.B., 1972, *The Serengeti Lion: A study of predator-prey relations*, Chicago: University of Chicago Press

——2007, *A Naturalist and Other Beasts: Tales from a life in the field*, San Francisco: Sierra Club Books

Schuette, P., Creel, S. & Christianson, D., 2013, 'Coexistence of African lions, livestock, and people in a landscape with variable human land use and seasonal movements', *Biological Conservation*, vol. 157, pp. 148–54

Shriner, D., Tekola-Ayele, F., Adeyemo, A. & Rotimi, C.N., 2018, 'Genetic ancestry of Hadza and Sandawe peoples reveals ancient population structure in Africa', *Genome Biology & Evolution*, vol. 10, no. 3, pp. 875–82

Silberbauer, G.B., 1981, *Hunter & Habitat in the Central Kalahari Desert*, Cambridge: Cambridge University Press

Spong, G., 2002, 'Space use in lions, *Panthera leo*, in the Selous Game Reserve: Social and ecological factors', *Behavioral Ecology and Sociobiology*, vol. 52, no. 4, pp. 303–7

Sokile, C.S., van Koppen, B. & Lankford, B., 2003, 'Ten years of the drying up of the Great Ruaha River: Institutional and legal responses to water shortages', *Environmental Science* <www.semanticscholar.org/paper/Ten-Years-of-the-drying-up-of-the-Great-Ruaha-and-Sokile-Koppen/8e2461b57d0f9b19cf49ef20745b80b8167d44a1>

Bibliography

Somerville, K., 2019, *Humans & Lions: Conflict, conservation & coexistence*, London: Routledge

Stander, P., 2009, 'Movement patterns and activity of desert-adapted lions in Namibia: GPS radio collars', Desert Lion Conservation Research Report

Stander, P.E., 1992, 'Foraging dynamics of lions in a semi-arid environment', *Canadian Journal of Zoology*, vol. 70, no. 1, pp. 8–13

Stander, P.E. & Albon, S.D., 1993, 'Hunting success of lions in a semi-arid environment', in N. Dunstone & M.L. Gorman (eds), *Symposium of the Zoological Society of London*, no. 65, pp. 127–43

Starkey, J., 2012, 'Families in fear as lions start eating out in the big city', *The Times*, 4 June

Stirke, D.W., 1922 (1969), *Barotseland: Eight years among the Barotse*, New York: Negro Universities Press

Sunquist, M. & Sunquist, F., 2002, *Wild Cats of the World*, Chicago: University of Chicago Press

Survival International, *Bushmen Aren't Forever—Botswana: diamonds in the Central Kalahari Game Reserve and the eviction of Bushmen*, 18 September 2006, London: Survival International

——2014, *Arrest and Beatings, Torture and Death: The persecution of Bushman hunters, Botswana, 1992–2014*, London: Survival International

Suzman, J., 2017, *Affluence without Abundance: The disappearing world of the Bushmen*, New York: Bloomsbury

Thomas, D.S.G. & Shaw, P.A., 1991 (2009), *The Kalahari Environment*, Cambridge: Cambridge University Press

Thomson, J., 1885, *Through Masai Land: A journey of exploration among the snowclad volcanic mountains and strange tribes of eastern equatorial Africa*, London: Sampson Low, Marston, Searle, & Rivington

Thruston, A.B., 1900, *African Incidents: Personal experiences in Egypt and Unyoro*, London: John Murray

Time Magazine, 1969, 'Anthropology: The original affluent society', 25 July, vol. 94, no. 4

Tucci, S. & Akey, J.M., 2016, 'A map of human wanderlust', *Nature*, vol. 538, no. 7624, pp. 179–80

Tucker, A., 2010, 'The truth about lions', *Smithsonian Magazine*, September

UNESCO World Heritage Committee, 2016, *Decision 40 COM 7A.47: Selous Game Reserve (United Republic of Tanzania)*, Paris, 15 November

Vaillant, J., 2010, *The Tiger: A true story of vengeance and survival*, New York: Vintage Books

Valeix, M., Fritz, H., Loveridge, A.J., Davidson, Z., Hunter, D.O., Murindagomo, F. & Macdonald, D.W., 2010, 'Does the risk of encountering lions influence African herbivore behaviour at waterholes?', *Behavioral Ecology and Sociobiology*, vol. 63, no. 10, pp. 1483–94

Valeix, M., Hemson, G., Loveridge, A.J., Mills, G. & Macdonald, D.W., 2012, 'Behavioural adjustments of a large carnivore to access secondary prey in a human-dominated landscape', *Journal of Applied Ecology*, vol. 49, no. 1, pp. 73–81

Valeix, M., Loveridge, A.J., Chamaillé-Jammes, S., Davidson, Z., Murindagomo, F., Fritz, H. & Macdonald, D.W., 2009, 'Behavioral adjustments of African herbivores to predation risk by lions: Spatiotemporal variations influence habitat use', *Ecology*, vol. 90, no. 1, pp. 23–30

Valeix, M., Loveridge, A.J., Davidson, Z., Madzikanda, H., Fritz, H. & Macdonald, D.W., 2010, 'How key habitat features influence large terrestrial carnivore movements: Waterholes and African lions in a semi-arid savanna of north-western Zimbabwe', *Landscape Ecology*, vol. 25, no. 3, pp. 337–51

Valeix, M., Loveridge, A.J. & Macdonald, D.W., 2012, 'Influence of prey dispersion on territory and group size of African lions: A test of the resource dispersion hypothesis', *Ecology*, vol. 93, no. 11, pp. 2490–6

van der Post, L., 1958 (2002), *The Lost World of the Kalahari*, London: Vintage Books

van Vuuren, J.H., Herrmann, E. & Funston, P.J., 2012, 'Lions in the Kgalagadi Transfrontier Park: Modelling the effect of human-caused mortality', *International Transactions in Operational Research*, vol. 12, no. 2, pp. 145–71

Bibliography

Verlinden, A., 1997, 'Human settlements and wildlife distribution in the southern Kalahari of Botswana', *Biological Conservation*, vol. 82, no. 2, pp. 129–36

Vidal, J., 2015, 'Craig Packer: "Cecil the lion's killer was unlucky and not altogether to blame"', *The Observer*, 4 October

Walsh, M., 2012, 'The not-so-Great Ruaha and hidden histories of an environmental panic in Tanzania', *Journal of Eastern African Studies*, vol. 6, no. 2, pp. 303–35

West, P.M., 'The lion's mane', *American Scientist*, vol. 93, pp. 226–35

West, P.M. & Packer, C., 2002, 'Sexual selection, temperature, and the lion's mane', *Science*, vol. 297, no. 5585, pp. 1339–43

Western, D. & Manzolillo-Nightingale, D.L., 2003, *Environmental Change and the Vulnerability of Pastoralists to Drought: A case study of the Maasai in Amboseli, Kenya*, Amboseli Conservation Program

Whitman, K.L., 2002, 'Safari hunting of lions: A review of policies, practices, and industry', in *Proceedings of the 2nd Meeting of the African Lion Working Group, Brandhof, 9–10 May 2002*, pp. 111–24.

——2010, *Pocket Guide to Aging Lions*, brochure, Metairie, LA: Conservation Force

Whitman, K., Starfield, A.M., Quadling, H.S. & Packer, C., 2004, 'Sustainable trophy hunting of African lions', *Nature*, vol. 428, no. 6979, pp. 175–7

Wilderness Safaris, 2010, *Kalahari Plains Camp—response to continued allegations*, July (press release)

Wildlife & Environment Zimbabwe, Matabeleland Branch, 2013, *Game Census for Hwange National Park and Surrounding Areas*, Bulawayo: Wildlife & Environment Zimbabwe

Wilson, V.J., 1997, 'Biodiversity of Hwange National Park—Part 1: Large Mammals and Carnivores', preliminary analysis report for Chipangali Wildlife Trust and Department of National Parks and Wildlife Management, Bulawayo: Chipangali Wildlife Trust

World Health Organization (WHO), Division of Vector Biology and Control, 1985, *Data Sheet on Pesticides No. 56—Carbofuran*, Geneva: WHO

Yeakel, J.B., Patterson, B.D., Fox-Dobbs, K., Okumura, M.M., Cerling, T.E., Moore, J.W., Koch, P.L. & Dominy, N.J., 2009, 'Cooperation and individuality among man-eating lions', *Proceedings of the National Academy of Sciences*, vol. 106, no. 45, pp. 1940–3

Young, A., 2009, 'Lion baiting: The use of agricultural poisons to kill wild carnivores in Africa', prepared for Panthera (unpublished briefing paper)

Zimmer, C., 2016 'A single migration from Africa populated the world, studies find', *New York Times*, 21 September

DVDs

BBC Natural History Unit, 2003, *Life of Mammals*, London

Brauer, H., 2009, *The Last Lioness*, Aquavision TV Productions, Johannesburg

Lion Conservation Programs

Lions may be in trouble, but innovative lion research and conservation projects are working to save lions from extinction across Africa. The following are among the better NGOs. Each can provide further information about the threats posed to lions and the work being done to save local lion populations. All accept public donations.

African Parks (africanparks.org)

This international NGO has taken on some of the continent's most troubled national parks and turned them into unlikely success stories. Under African Parks' management, anti-poaching measures operate alongside community engagement programs, habitat restoration and the restocking of wildlife populations. Liuwa Plain National Park in Zambia (Chapter 3) is one of seventeen parks managed by African Parks.

Big Life Foundation (biglife.org)

Run by veteran conservationist Richard Bonham, Big Life operates anti-poaching militia in Kenya, targeting and arresting poachers of lions, rhinos, elephants and other wildlife in Amboseli, the Chyulu Hills and across the border into Tanzania.

Ewaso Lions (ewasolions.org)

This program is similar to the Lion Guardians idea, but works among the Samburu people of northern Kenya. In addition to protecting lions and livestock, Ewaso Lions aims to build strong links between lion conservation and Samburu communities. It works on the community conservancies surrounding Samburu National Reserve and is headed by Dr Shivani Bhalla.

Lion Guardians (lionguardians.org)

The Lion Guardians of Amboseli pioneered the adaptation of local cultural traditions among the Maasai to empower young Maasai warriors—like Meiteranga Kamunu Saitoti (Chapter 1)—who once killed lions to protect them. The project continues with great success and has been taken on, with local variations, in communal areas around Hwange National Park in Zimbabwe, Ruaha National Park in Tanzania and elsewhere.

Lion Recovery Fund (lionrecoveryfund.org)

Taking the crisis facing many lion populations across Africa as its starting point, the well-connected Lion Recovery Fund makes grants to conservation programs in an attempt to turn the tide and save lions from extinction. It supports a wide range of research and practical initiatives, and has funded projects run by Panthera, WildCRU, Big Life Foundation, Ewaso Lions, the Ruaha Carnivore Project and others.

Maasai Wilderness Conservation Trust (maasaiwilderness.org)

Close to where Kenya's Amboseli Basin meets Tsavo West National Park, this trust works on health, deforestation and education programs among Maasai communities, as well as lion conservation programs. These include the Simba Scouts (using local Maasai rangers to protect lions) and Wildlife Pays (through which local communities receive money for the wildlife that lives on their land).

Panthera (panthera.org)

A world leader when it comes to big-cat conservation, Panthera funds and operates numerous lion conservation programs across Africa under the guidance of Dr Paul Funston. Their Project Leonardo operates in a number of African countries, and involves a mixture of scientific study and on-the-ground community engagement.

Soft Foot Alliance (softfootalliance.org)

The brainchild of Brent Stapelkamp (Chapter 2), formerly of the Hwange Lion Research Project, and Laurie Simpson, the Soft Foot

Alliance uses everything from permaculture to mobile *bomas* in trying to minimise conflict between people and lions (and other wildlife) around Hwange National Park in Zimbabwe.

WildCRU (Wildlife Conservation Research Unit; wildcru.org)

Based at Oxford University, WildCRU combines long-running wildlife research programs with in-situ conservation programs. Among their projects are the Trans-Kalahari Predator Programme (including the Hwange Lion Research Project in Zimbabwe; Chapter 2), the Laikipia Predator Project in Kenya, the Ruaha Carnivore Project in Tanzania, and lion studies in West and Central Africa.

Wildlife Conservation Society (wcs.org)

One of the largest wildlife conservation NGOs on the planet, WCS has a growing big-cat program under the leadership of Dr Luke Hunter. Lions are a major focus, particularly implementing anti-poaching patrols designed to halt the killing of lions and their prey in some of Africa's most imperilled protected areas.

About the Author

Anthony Ham is one of Australia's most experienced nature and travel writers. For more than two decades, he has been travelling to the earth's wild places in search of stories, to Africa, the Amazon, the Arctic, and Outback Australia. His work has appeared in *The Age*, The Sydney Morning Herald, The Monthly, *Virginia Quarterly Review (VQR)*, *National Geographic Traveler*, *BBC Wildlife*, *Lonely Planet Traveller*, *Africa Geographic*, *The Independent*, *Travel Africa*, and elsewhere. Through his writing, Anthony's readers have observed up close Africa's most endangered elephant herd, travelled to remote villages behind al-Qaeda lines, and experienced the beginnings of Libya's long descent into chaos. Anthony has also written or co-written more than 135 travel guides for Lonely Planet, including the bestselling guides *Kenya, Tanzania, East Africa, Southern Africa, Botswana & Namibia*. He believes in the power of the written word, the enduring power of stories, and the importance of writing beautifully about important things. He lives in Melbourne with his wife and two children.

www.anthonyham.com